The International Guide to Wood Selection

Interior of a house in Bromborough, Cheshire. Architects: Nelson and Parker

The International Guide to Wood Selection

Edited by Ezra Levin

DRAKE PUBLISHERS INC NEW YORK

ISBN 87749-222-0

Published in 1972 by
Drake Publishers Inc
381 Park Avenue South
New York, N.Y. 10016

Printed in Great Britain

Contents

Introduction

Although wood has been handled by mankind since the beginning of time some builders and architects may be less familiar with many of its specific properties than they are with other building materials and, a great deal is often left to chance in specifying and detailing. This is especially so to-day in view of the many more timbers now imported from semi-tropical and tropical countries.

This publication provides a great deal of information on the various properties of many species of different woods and should prove to be a useful guide.

Contrary to common belief, wood can be highly weather resistant and many species are extremely durable even if left untreated. This is shown by historic timber framed buildings and by ancient wood carvings and panels some centuries old.

Wood has excellent insulating properties. Its low thermal conductivity coupled with its comparatively low weight results in such a low thermal capacity that a room having its walls lined with it responds quickly to increases in temperature of the air and thus provides a sense of comfort seldom equalled. Wood is warm to the touch and little body heat is lost to it by radiation. The same applies to timber floors and the feeling of comfort is enhanced here by the resilience of the surface. Wooden windows, especially if combined with double glazing, are free from condensation often associated with metal frames.

This low thermal conductivity combined with a known and regular rate of burning of wood, can, with intelligent design, provide effective fire barriers. Timber is classed as "not easily ignitable" and, although it burns it can provide excellent fire resistance.

Acoustically, wood finds many applications. Qualities of resonance and adaptability make timber panels of different kinds most valuable tools in the tuning of concert halls and sound studios. The flexibility of partition designs based upon studded frameworks makes them universally acceptable. Such partitions can achieve widely different degrees of sound insulation and/or absorptive properties to meet the most exacting requirements.

Its easy workability by the tools of the handyman – as well as by the most sophisticated production machinery – is well known. The flexibility in component design it makes possible has resulted in a widening acceptance by industrial building producers and it could well become the principal material in one, two and three story housing in the future. Wood, as the raw material for particle boards and laminated components will increase its versatility and adoption.

Lastly, the wealth of textures and colours of the multitude of different species and different conversions for use provides a range of natural beauty unequalled in any other single material. This provides the desired scope for designers who to-day look to their materials to provide aesthetic qualities, rather than to applied ornament.

Interior of a house in Bromborough, Cheshire. Architects: Nelson and Parker

Section one: the material

Basic properties of wood

Definition of hardwoods and softwoods

Commercial timbers are divided into two types: 'hardwoods' and 'softwoods'. Hardwoods are produced from broad leaf trees and softwoods from coniferous trees. Most commercial hardwoods are harder than the majority of commercial softwoods.

Structure of wood

A brief description of the structure of wood is desirable so that the terms used later on may be understandable.

Wood, like all plant material, is cellular. The cells (or 'wood elements') consist, for the most part, of microscopically small fibres but other types of cell also occur. Most cells, including the fibres, are orientated in the direction of the axis of the trunk; one exception is the tissue known as 'rays' (sometimes, but inaccurately, termed 'medullary rays'). Rays are plates of cells running in a radial direction (see figure 1). When the rays are very thick (as in oak) they produce a characteristic figure in timber when it is quarter-sawn.

In temperate countries a tree produces a new layer of wood just below the bark every growing season. Growth ceases during the winter months. This process results in clearly defined concentric rings of wood being observable on the ends of logs (see figure 2). Such rings are called 'growth rings' or 'annual rings', and give rise to various markings on sawn timber or veneers.

In the tropics there is no winter period when cold prevents tree growth, and therefore most tropical timbers do not show growth rings.

Knots are a familiar feature of the structure of wood. A knot is a portion of a branch embedded by the natural growth of the tree and normally originates in the centre of the trunk or branch.

Sapwood and heartwood

In both hardwood and softwood logs there is an annular band of wood nearest to the bark, called the 'sapwood'. The central core of the wood inside this band

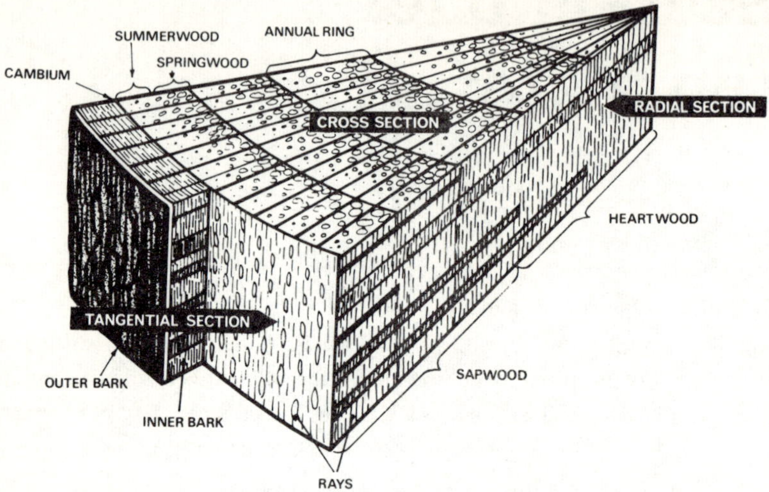

Figure 1

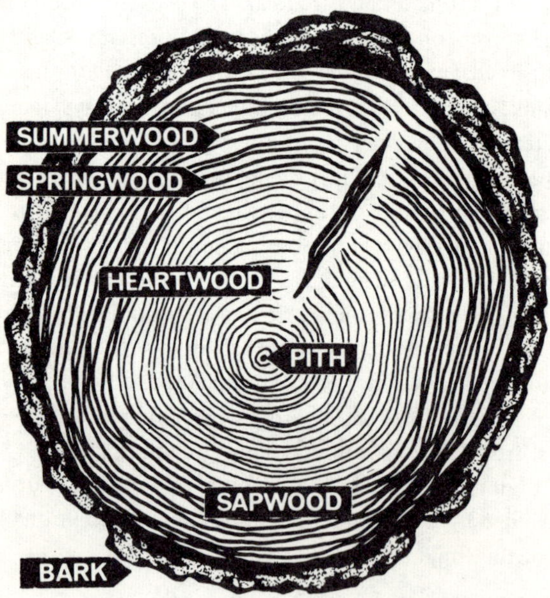

Figure 2. Cross section of softwood log showing bark, wood and pith

is known as the 'heartwood'. The function of the sapwood is to transport the sap from the roots up the trunk to the leaves. The heartwood does not carry out this function, but gives mechanical support to the trunk. Very often the heartwood is darker in colour than the sapwood and has a more decorative appearance when sawn or cut into veneer. The sapwood generally is about the same weight and has the same strength properties as heartwood, but it is usually more prone to attack by fungi and insects. To compensate, as it were, for its lack of durability the sapwood is normally more absorbent and so easier to treat with preservatives.

The width of the sapwood band varies with species from an inch or less to about seven inches.

The colour of heartwood varies from white to black, with reds, yellows and browns and a wide mixture of these colours in between. Some variation in colour occurs in different logs of the same species. Colour changes also occur after the wood is cut and exposed to light and air; visually the colour deepens but may change more radically, e.g. iroko often turns from yellow to dark brown, Rhodesian teak changes from pale brown to a deep red. The process of steaming or even kilning may change colour; for instance, the natural whitish colour of beech becomes pink on steaming. A classification of timbers according to colour is given in the table in Appendix A on page 85.

Grain, texture and figure

When describing timber, the terms 'grain', 'texture' and 'figure' are used – and frequently misused. The British Standard 565: 1949 ('Glossary of terms applicable to timber, plywood and joinery') defines these terms as follows:

grain: 'a term referring to the general direction or arrangement of the fibres and other wood elements' (or cells). The Standard also includes the following note: 'Sometimes used to describe the various structural features of timber, e.g. cross grain, silver grain; also the plane of the cut surface of timber, e.g. end grain, side grain; the nature of the worked surface of timber, e.g. chipped grain, raised grain; the type of fracture, e.g. short grain'.

texture: 'The distribution and relative size of the wood elements', thus there are coarse- and fine-textured woods, etc.

figure: 'The ornamental markings, seen on the surface of manufactured timber, formed by the structural features of the wood or by variations in the natural colour'.

Timber

Softwoods

Most commercial softwoods used in this country are fairly soft and light (weighing about 400-550 kg/m^3 when seasoned). They are generally easy to work with both machine and hand tools and so suitable for most general utility purposes. The majority of softwoods are fairly stable in varying conditions of humidity – small to medium movement categories.

Softwoods, depending upon the grade, may be used as a decorative medium – growth characteristics such as knots can also be attractive. The colours range from an off-white to a reddish brown. Flat-sawn timber has a more attractive figure than quarter-sawn.

Hardwoods

The working qualities of hardwoods vary considerably from mild and easy to very difficult, largely dependent upon their density. Modern cutting tools largely overcome 'difficult' processing but the cutting angle may sometimes have to be adjusted.

Many hardwoods are stable and fortunately only a small proportion have large movement values. It is emphasised that this property is directly related to the moisture content and atmospheric conditions existing in a building (see section on moisture content below). Hardwoods give a far wider range of colour and grain effects than do softwoods.

It must be remembered that as timber is a natural product colour variations can be considerable within a single piece and this in turn has to be considered in the overall decorative effect desired.

Conversion of timber

'Conversion of timber' is a term generally used to mean the cutting up of a log into sawn timber; it may also be used to refer to cutting a log into veneer. In the conversion to sawn timber, various types of saw are used, principally circular saws, band saws and frame saws, the latter are more common in Scandinavia and Canada than in this country. Logs may be cut in various ways, for example 'through and through' (see figure 3) or 'quarter-or rift-sawn' (see figure 4). In the former method, most of the boards will be 'flat sawn' ('plain sawn'), which means that the growth rings meet the face of the board in any part of an angle of less than 45°. In quarter-sawn timber the growth rings meet the face at an angle of not less than 45°. When this angle is not less than about 80° the terms 'edge sawn', 'edge grain' or 'fully quarter-sawn' are sometimes used (although not accepted as standard terms in BS 565).

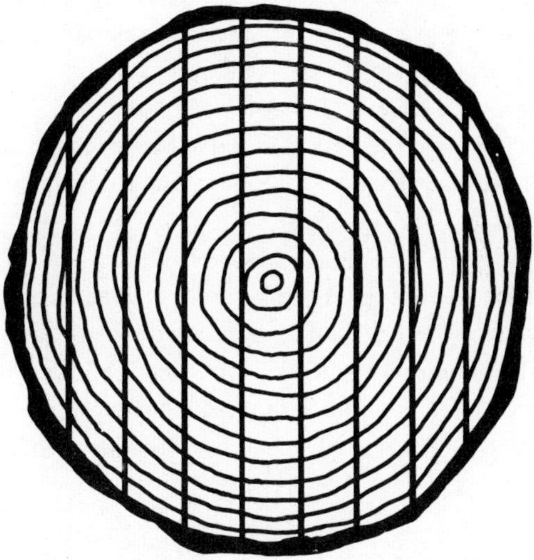

Figure 3. Through and through cut

Figure 4. Quarter sawing

The method of conversion has a bearing upon the resultant figure or pattern obtainable. Flat-sawn timber from softwoods has a more decorative appearance than quarter-sawn, but this does not generally apply to hardwoods since their figure depends largely upon grain irregularities as well as natural features, such as rays.

Quarter-sawn boards shrink less than flat-sawn material and generally give more even wear, as, for example, in flooring. (See paragraph on moisture movement opposite).

The inherent variability of timber has led to the study of natural features exposed on conversion, and the formulation of grading rules in which the frequency and size of these natural features are defined. The highest grades are those with the minimum amount of variation, the lowest those with the maximum number commensurate with reasonable saleability and where appearance or strength is a secondary consideration. See BS 112.

Seasoning

The wood of all trees when felled contains a considerable amount of moisture; this may amount to 150% or more of the oven dry weight (i.e. the weight of the wood after it has been dried at 100°C in a ventilated oven until no further loss of weight occurs). Timber is of very limited use in the damp condition because it will shrink, check and warp as it dries, it is difficult to work and to paint or polish, it is less strong and durable than when dry, and it is difficult to treat with a preservative, if this is desirable. It is, therefore, necessary to dry timber so that it will be in equilibrium with the humidity conditions of its ultimate use.

This process is known as 'seasoning'. There are two main methods of seasoning in current use, viz., air seasoning and kiln drying. In the climatic conditions obtaining in this country, it is not possible to air season to a degree required in many modern service conditions, particularly for woodwork in centrally heated buildings. Kiln drying makes it possible to control the natural forces required for air seasoning, i.e. moisture, heat and ventilation and so to reduce the moisture content of timber to almost any degree.

It is not enough merely to specify 'kiln dried' timber, as very brief kilning is sometimes done to dry the surface only (to prevent staining during shipment). The moisture content required should always be specified (see p.65).

Moisture content

Wood is a hygroscopic material and so reacts to changes in humidity of its environment. If it is dried to a condition where it is as near as possible in equilibrium with its surroundings, the amount of shrinkage and swelling associated with changes in moisture content will be reduced to a minimum.

It is desirable to specify a moisture content (or a range of moisture contents) for timber to be used in most parts of buildings. With such constructional timbers as rafters and joists, moisture content can be fairly high when the wood is installed, as it will dry out *in situ* without the resultant shrinkage causing any serious trouble. Interior joinery, on the other hand, is a case where correct moisture content is important if ugly open joints, splits or worse troubles are to be avoided. This is particularly true in centrally heated buildings where the moisture content should not exceed 10%. For normal heating in private houses 14% would prove satisfactory for joinery work. For structural members and carcassing timber a moisture content of up to 22% is acceptable. For timber which is to be painted or varnished moisture content should not exceed 18%. See also CP 112 about moisture content.

Moisture movement

Although timber may have been dried to the correct moisture content for its use, there are likely to be small dimensional changes (known as 'movement' or 'working') with variations of atmospheric conditions. The user of timber is generally more concerned with movement than with shrinkage of wood on drying from the green to the seasoned state. Tests have been carried out to determine the amount of movement in a large number of timbers. Wood specimens are subjected to a relative humidity of 90% and then to 60% (both at 25°C) for considerable periods and the resultant amount of shrinkage measured. Based on the results of these tests most commercial timbers have been classified into timbers with small, medium and large movement, and this information is included in the table in Appendix A (see also Forest Products Research Laboratory Leaflet No. 47 'Movement of Timbers').

There are appropriate Codes of Practice for floors, doors, windows and structural uses of timber in building which should be consulted. The ideal material to apply to timber for the prevention of moisture pick-up has not yet been discovered, but aluminium primer of a leafing quality has been proved efficient in this respect.

As the end grain of timber is far more absorbent than the side grain, it is an excellent plan to paint every possible end grain surface with aluminium primer. Back-priming of woodwork adjacent to brick walls, etc, with the same type of primer is also a wise precaution.

Decorative veneers

Veneers are thin sheets of wood produced for decorative purposes and are cut in a manner to expose the natural beauty of the wood to its best advantage.

Conversion

Decorative effects are obtained by cutting the log to reveal its colour, grain and figure. Rotary cut veneers are produced by bringing a knife to bear against the full length of a log in a lathe, and veneers so produced are cut tangentially to the growth rings of the log. Sliced veneers are cut by a long knife moving across the face of a log or flitch (part of a log) removing a sheet of veneer at each forward stroke. In some machines the wood is moved across the knife. Most decorative veneers are sliced.

Decorative effect

The decorative effects obtained will depend on:

1 Colour
2 Figure
3 Matching arrangement of veneers.

How these three factors are combined will depend on the effect desired. Types of figure will be found listed under Section 6, Specification of Veneers.

Advantages of veneering

1 **Range of effect**: Veneers are manufactured from the cream of the world's hardwoods and from a few selected softwoods such as yew, pine and larch.

This tremendous range of colour, grain and figure provides the architect or designer with an almost endless variety to choose from, a variety which can provide for all purposes.

2 **Economical use of material**: Veneer enables rare and expensive woods to be supplied at acceptable prices. While the less expensive woods are highly competitive with any other form of decorative facing, veneers of all types are more advantageous than might appear from comparisons of first cost because when laid and polished they have another great asset – little or no need of maintenance.

3 **Workability**: Used in veneer form, even the most refractory timbers become tractable, and, applied to a properly constructed panel, give no trouble

in the most exacting conditions – e.g. in a building with central heating. Questions of dryness and seasoning do not arise with veneer; they are ready for immediate use.

Veneered panels

Regardless of method of conversion from the solid to the veneer form, the veneer can subsequently be glued to a core of cheaper material to form a high grade panel. This core should be a flat, stable surface and may be blockboard, laminboard, hardboard, plywood, particle board or solid timber.

The choice of adhesive will be decided by the nature of the veneers to be laid, the value of the finished product, liability to stain and other conditions. Mainly synthetic resin adhesives such as urea formaldehyde in conjunction with a hardener and cereal extender are preferred, but casein mixtures, animal and skin glues are all used.

The decorative veneer should be laid at right-angles to the grain of the face of the core whether based on commercial plywood, blockboard, laminboard or a prepared base. Undulations of the core should be smoothed out by sanding or scraping and the moisture content of the board and veneer matched to reduce differential shrinkage.

The finished board should be of balanced construction, i.e. a construction such that the forces induced by changes in moisture content will not cause warping. In practice this means that any pair of corresponding veneers or layers on either side of the centre line are of the same species and thickness, and are laid with the grain in the same direction as each other.

Balanced construction

Where a standard veneered board is purchased then this factor will have been taken into account by the manufacturer. In cores veneered to order, a compensating veneer will have to be fixed to the reverse face. This compensating veneer is usually of low quality and of the same species or of a species of similar properties, grain direction, and thickness, to the face veneer.

On occasions when a highly decorative veneer with a complicated figure is desired then it is best to use an underlay veneer between the face of the core stock and the decorative finish. The underlay should be a cheap, non-decorative veneer which is stable. This prevents any disturbance of the comparatively inelastic decorative veneer, should changes in atmospheric conditions cause a slight surface movement of the core.

Plywood

General properties

Plywood by virtue of its method of construction exhibits well defined characteristics. Some of these are specific to different types and countries of origin which are enumerated in Appendix 'E'. However there are many general qualities which are summarised here:

1 **High uniform strength.** Wood is 25 to 40 times as strong along the grain than across it. Crossing of the grain at right angles tends to equalise the strength in both directions. This uniformity in strength increases as the number of panels increases. This also contributes to its great strength when loaded edge-wise which gives considerable resistance to buckling. External grade plywoods available from most sources are a suitable cladding material and when used as sheathing eliminate the need for 'cross bracing' (see p.41 on 'cladding') and also contribute to thermal and sound insulation. (See Section 2). There are many other structural uses of plywood which will be discussed in the appropriate publication.

2 **Reduction of deformation and moisture movement.** Solid timber exhibits considerable movement across the grain, but generally only a negligible amount in longitudinal direction. The balanced construction of plywood tends to equalise these stresses making moisture movement generally unimportant. It is in the order of 0.01% per 1% change in moisture contents. For example a sheet 1500 mm x 3000 mm will change not more than 2.5 mm over a range of 5% change in moisture content. Tolerances are in the order of 0.1%.

3 **Non-splitting.** Due to its cross grain construction plywood has no cleavage. It cannot split. Nails, screws and rivets can therefore be placed in close proximity to the edge of the board. In addition, the criss cross arrangement of the fibres offers considerable resistance to the pulling through of nails or screw heads.

4 **Other qualities.** Plywood being a manufactured material can be obtained in large size sheets which facilitate ease of erection and fixing. Standard panels are usually available measuring 2400 x 1200 mm with thicknesses ranging from 6.5 – 32 mm. But larger unbroken sheets can be obtained scarf or finger jointed. Although it is usually used in flat sheets the resilient properties of plywood enable it to be bent to a radius of curvature proportionate to its thickness and construction without suffering damage. It is easily sawn into simple or complicated shapes and apart from the resulting uses as a finishing or decorating material these qualities make it a very suitable material for the formation of acoustic panels, wood having excellent

resonant properties (see Section 2 p.29). Plywood is easy to finish since no planning is required, and it may be obtained ready sanded. It may be polished, varnished, stained, painted, etched and veneered.

Care and use of plywood

Moisture retardance. When large plywood panels are used in situations where there is likely to be pick-up and loss of moisture then it is best to have the edges and sides of the boards sealed. This may be done with a sealer or with a coating of the same type as the finish. This is of major importance when the material is used externally. It should be noted that a surface application is not sufficient to withstand continuous exposure to the atmosphere. The face should first be primed with a suitable primer, and the paint manufacturers' instructions followed thereafter for the best results. Edges of the boards may be sealed with paint and/or lead paste to prevent the infiltration of moisture which could result in the breakdown of the finish. The backs of the boards, even though not necessarily subject to direct weather, should be treated with a red or white lead if used under damp conditions.

Durability. The term 'weatherproof' when applied to plywood refers only to the adhesive used in its manufacture. No delamination of the plies will occur due to dampness or exposure to weather in plywood of exterior grade WBP but the timber from which the plywood is made up may nevertheless be liable to attack by a wood destroying agent if it is of a non-durable species. Breakdown of the timber by such wood destroying agencies can be prevented by:

1 The use of plywood constructed throughout with a species known to be durable

2 Affording a suitable protection to the face, back and edges of the material

3 Treating the plywood with a suitable preservative.

Where conditions are conducive to decay, the advice of the manufacturer or merchant should be sought in order that a suitable material is obtained for the work in hand.

The adhesive types available will be dependent on the country of origin but generally speaking British made and many imported plywoods have the following designations and characteristics in respect of quality and bonding:

INT – Interior. Adhesives of this group give a strong durable bond under dry conditions with limited resistance to moisture. They are not required to withstand attack by micro-organisms.

MR – Moisture and moderately weather resistant. To this group belong those adhesives which will survive full exposure to weather for limited periods, but fail under the boiling water-test. MR adhesives are highly resistant to micro-organisms.

BR – Boil resistant. These adhesives have a good resistance to weather and to the boiling water test, but they fail under prolonged condition of full exposure to weather for which WBP adhesives are satisfactory. They will withstand cold water for many years and are highly resistant to micro-organism attack.

WBP - Weather and boilproof. Adhesives that by systematic tests and their record in service have proved highly resistant to weather, micro-organisms, cold and boiling water, steam and dry heat. Such glues are more durable than the wood itself.

Further details of adhesives for plywood manufacture in the United Kingdom will be found in BS 1203: 1954 'Synthetic Resin Adhesives for Plywood' and BS 1455: 1956 'British Made Plywood for General Purposes'.

Adhesives of plywood imported into the U.K. in some cases conform to British Standard 1203: 1954, but this is not so for all manufacturing countries. It is still common practice amongst certain foreign manufacturers to designate adhesive types by the old British Standard values of A70 and AX100. Whilst A70 has a similar moisture resistance to MR, it should be noted that plywood bonded to AX100 is not comparable with WBP, and should not be used under conditions of exterior exposure. It is therefore suggested that when there is doubt as to the actual bond quality the supplier of the plywood should be consulted.

Particle board

General properties

Particle board is made from specially prepared particles of ligno-cellulosic materials such as wood, flax, or hemp, bonded with synthetic resins or other organic binders. All particle boards generally conform to the requirements of BS 2604: 'Resin-bonded Wood Chipboard' in which are given requirements for dimensions and tolerances, freedom from foreign particles, moisture content, thermal conductivity, surface spread of flame and minimum strength properties. Test and test methods for wood chipboard and other particle boards are specified in BS 1811: which also incorporates an addendum relating to flooring grade boards.

Particle board is manufactured to highly developed mechanical techniques which ensure strict regularity of size, thickness, moisture content and other physical characteristics. Thus large flat panels are produced which have no direction of grain and are free from defects.

Application

Particle boards can be used for a wide variety of applications such as roof decking, flooring, wall lining, partitioning, built-in fitments and other joinery such as door cores. Within these applications it is possible to obtain boards of specific nature and characteristics e.g. flooring grade boards, ready-to-paint grades, boards faced with wood, plastic and hardboard veneer, low-density panels (usually flaxboard) for roof-decking etc. For new building, particularly industrialised techniques, architects are generally recommended to consider the use of particle board early in their designs in order to obtain full benefit from the material.

When particle boards are used with thermal insulation in mind the 'K' value is similar to that of most softwoods i.e.

for wood chipboards	0.144 W/m°C
for flat particle boards	0.072 " "

Storage on site

What has been said about storage on site for joinery etc. on p.36 is of equally vital importance for particle board. In any case it should not be left on site too long prior to use and should be stored flat.

Costs and types

Particle board being a manufactured material is not subject to grading for natural defects carried out for plywood or solid timber. It is obtainable in varying densities which affect strength and thermal properties and determine the purpose for which it is used. Generally speaking on a cost per square metre it is cheaper than other comparable materials.

Particle boards are manufactured in many countries including the U.K. For imported boards the following countries are the principal sources of supply:

Wood chipboard

Bulgaria	Irish Republic	Portugal
Czechoslovakia	Israel	Rumania
Finland	Norway	Soviet Union
W. Germany	Poland	Sweden

Flax board

Belgium	France
Czechoslovakia	Poland

Care and use of particle board

Where desired, butt jointing of the edges may be readily carried out, but it is essential that these be liberally coated with adhesive preferably with a synthetic resin such as urea formaldehyde. Flat pressed boards may be readily butt jointed in any direction, it being immaterial whether or not adjoining edges are cut in the same direction. However, this does not apply to extruded wood chipboards where adjoining edges must match. With all boards sufficient joint strength is attained without use of a tongue and groove or a loose tongue, although these may be necessary for location purposes. When protecting particle board against moisture penetration, it is again essential that finishing treatment be applied to face, back and edges.

Blockboard and laminboard

Description

Blockboard and laminboard are essentially similar products to plywood but in practice the minimum thickness is 12.5 mm. In blockboard the construction consists of strips up to 25 mm wide placed together with or without glue between each strip to form a slab. This core is sandwiched between outer veneers with the grain direction at right-angles to that of the strips. In some constructions there may be double outer plies on each side of the core, but the grain of all the veneers runs at right-angles to that of the core. These types are generally known as three-ply construction.

When the length exceeds the width, the blockboard should be five-ply construction, i.e. a core of the same construction as mentioned above with a veneer on each side with the grain running at right-angles to that of the core, and an outer thinner ply with the grain direction parallel to that of the core.

In laminboard the core is built-up from strips of wood or veneer 3 – 7 mm wide which are glued together to form a slab. This is glued between outer veneers which have the grain direction at right-angles to that of the core strips as in blockboard. See drawing, p.74.

General properties

Such constructions do not have the uniform strength that characterises the veneer construction of plywood. They do however possess practically similar dimensional stability, laminboard being the more stable of the two.

Blockboard and laminboard permit jointing as desired. Edges of the boards can be moulded or veneered without developing a stripe effect which becomes visible when a veneer construction is so handled. For fine work the core is banded on two sides or along all four edges with timber of the same species as the face veneer.

Laminboard is heavier than blockboard since more glue, and, in some cases, denser timbers are used in its construction. Its greater cost limits it to high class work.

Care and use of blockboard and laminboard

Blockboards and laminboards are not normally manufactured for exterior use. The adhesives used are normally of **INT** or **MR** quality. In many cases the timbers used for the core are non-durable. It is therefore essential that whenever such constructions are liable to be exposed to damp every precaution is taken to prevent penetration of moisture into the interior. This should be carried out as detailed for plywood.

Acoustic panels in London's Royal Festival Hall. The panels measure approximately 24 m x 12 m, are of stressed-skin construction, and are the largest in Europe

Section two: performance data of timber

Apart from its structural properties three basic qualities of timber already referred to in the introduction need further explanation.

Thermal insulation

The thermal conductivities for timber are favourable compared with other materials used in building. They vary slightly with different densities, moisture content and species, the lower densities having somewhat lower conductivities. In addition to its low conductivity timber possesses low thermal capacity which means that it will warm up rapidly when exposed to sources of heat. This makes it very suitable in situations of intermittent heating. Its rapid response to warming up also reduces the risk of surface condensation already referred to in connection with window frames. The following list gives the thermal conductivity values of some commonly used timbers together with their densities at 15% moisture content. A more complete list can be found in the TRADA Design Guide.

Western Red Cedar	0.77 W/m°C	338 kg/m³
Douglas Fir	1.00 "	528 "
Iroko	1.40 "	640 "
Mahogany	1.65 "	705 "
Parana Pine	1.12 "	545 "
Pitch Pine	1.65 "	690 "

An all timber stud wall has a total 'U' value from 2½ to 3 times better than a brick cavity wall.

Acoustic properties

Sound absorption

Wood is the most frequently used material for platforms in auditoria. For wall and ceiling application its versatility for both aesthetic and acoustic purposes is unsurpassed. Acoustically when used in various forms in conjunction with other complementary absorptive materials it can provide optimum absorption throughout the audible range. At the same time decorative scope is limitless and durability (resistance to damage) an inherent feature. In the solid panel form when mounted over a cavity it can be tuned to achieve the desired low frequency absorption. In perforated or slatted configuration it can achieve any medium or high frequency absorptive properties required. All film sound stages and television studios employ wooden frames for accommodating the absorptive treatment and many concert halls employ wood panelling for the dual purpose of decorative and acoustic treatment, the advantage of which is that the acoustic treatment can be modified at will if, after construction, the reverberation time is required to be changed. The Royal Festival Hall and more recently the Queen Elizabeth Hall are good examples of this.

Sound insulation

Sound absorption and sound insulation are commonly confused. Absorption is the ability to reduce reflections, whereas insulation is the quality of sound resistance offered by a barrier/partition. In general, we think of walls and/or floors when considering sound insulation in buildings or compartments of buildings. It is accepted that, for simple constructions, mass (weight per unit area) provides the most effective barrier to airborne sound. However, it is not always practicable to incorporate massive structures in buildings – either because of their dead load and the difficulties of supporting them or their permanence. Economy may also be a consideration and speed of erection can be important. Good sound insulation can be achieved with lighter weight structures – if adequate thought is given to detailing and, where necessary, to discontinuous construction.

Timber is a versatile material and common forms of construction involve studs and lining materials behind which complementary sound resistant materials may be incorporated unseen. Timber framed party walls in semi-detached and terrace type domestic housing are now capable of meeting the highest standards. Timber studded partitions for offices can be designed to obtain any degree of sound insulation required – from the barest minimum in the case of low cost ones separating adjoining general offices or those between general offices and corridors, to the most effective sound resisting types separating the higher executive offices. In these latter cases discontinuous forms of construction (parallel but unconnected sets of studs) are used. Knowledgeable design and attention to detail can result in a very high sound insulation within minimum overall thickness.

Timber floors can be designed to meet Building Regulation requirements and in these cases discontinuity is achieved by means of floating rafts isolated by means of mineral fibre mats. Even where concrete sub-floors are used it is common practice to top them with a floating wood raft for impact sound resistance and for warmth and for general improved appearance.

The big attraction of wood in these circumstances is that good acoustic results can be achieved whilst maintaining flexibility of layout and the natural beauty and warmth associated with this material.

Fire resistance

The Building Regulations 1965 opened up a number of new opportunities for the use of timber both as a structural material and for finishes. 'Fire Resistance' is determined by tests detailed in BS 476: Part 1 and is required for all 'elements of structure' listed in the Regulations. Roof structures and non-loadbearing partitions are **not** included under 'elements of structure'. Non-loadbearing parts of external walls are excluded from fire resistance requirements, provided they are not too close to a boundary.

Where very light timbers are used, say 50 mm x 75 mm studs, a protective lining of asbestos, plasterboard or other suitable material may have to be provided to obtain a half hour fire resistance. On the other hand, where large timber sections are used structurally they may have sufficient 'sacrificial' timber to provide the necessary fire resistance. The approach to fire resistance here is based on the fact that timber usually chars at a predictable rate of 0.6 mm per minute. (Certain hardwoods such as teak have an even slower rate of about 0.5 mm per minute). The structural uses of timber with regard to fire are dealt with in more detail in other TRADA publications. (See bibliography). Here we are more concerned with non-structural uses of timber.

The Building Regulations in this context deal mainly with 'spread of flame'. In this respect materials are divided into classes, class 0 being equivalent to non-combustibility, classes 1 and 3 being increasingly combustible. Timber in its untreated state falls generally in class 3, as far as flame spread is concerned. Certain timbers, like Western Red Cedar, with a density of less than 400 kg/m^3 fall into class 4. But all timbers can be treated to give a class 1 performance by impregnation with fire retardant chemicals or by surface coatings including clear ones. Class 0 boards can be produced by having very thin timber veneers on non-combustible backings.

The general requirement is for class 1, but notable exceptions, on the lenient side, are that the ceilings of 'small residential buildings' (one and two-storey houses) and all the surfaces of 'small' rooms (not larger than 30 sq m (300 sq ft) in floor area, in the case of non-residential buildings, or 4 sq m

(40 sq ft) in the case of residential ones) may be of class 3 material. A 'ration' of class 3 material is also allowed on the walls of any room (e.g. as fixed panelling) up to an area equivalent to half the floor area of the room, with upper limits of 20 sq m (200 sq ft) in the case of residential buildings and 60 sq m (600 sq ft) in the case of non-residential buildings. Exceptions, on the severe side, are that wall and ceiling surfaces of rooms in 'Institutional' buildings for the sick, disabled, aged or very young and circulation spaces in all buildings, except 'one and two-storey residential' ones, must comply with a class 0 specification.

Exterior walls may be clad with timber – the distance of the wall from the boundary being controlled by the rule of 'unprotected area' detailed in the Building Regulations.

The only significant limitation to the use of timber in roofs applies to cedar shingles. This refers to the material's resistance to an external fire and the Building Regulations stipulate a minimum distance from the boundary of 12 m. Otherwise timber may be freely used to support other types of roofing.

Section three: design and fixing

The advantages of timber for non-structural or semi-structural uses are outlined in the Introduction. These advantages are, however, accompanied by a responsibility for careful design, specification and workmanship. It is outside the scope of this publication to consider all the design problems involved, or to make a detailed examination of specification clauses regarding material and workmanship, but attention is drawn to the following points of particular importance to ensure satisfactory appearance and adequate performance.

Design

The choice of shape and sizes of moulded sections whether in built-up frameworks such as window and door frames, curtain walling, etc. or as weatherboards, architraves, cover moulds, etc. is largely based on the designer's wish to obtain a particular aesthetic expression or effect. However, such practical matters as strength, weathering, rebating, etc. as well as economics must be taken into consideration. It is obviously undesirable to demand special and intricate patterns, requiring the setting up of machines and the grinding of special cutters, when economy demands the use of standard components or sections. Avoid designing undercuts. Where large quantities are involved, however, special machining may become an economic procedure.

Sizes must also be related to the nominal sizes of commonly available stocks since these latter vary with species and source of supply. Detailed information on this point is contained in Appendices 'B' and 'C' Generally it is better and also more economical to use stock sizes than to rely on conversion down to the required dimensions.

Care must also be taken when selecting a species that attention is given to their working and other properties having regard to the kind of machining and other operations to be performed both in the shop and on the site. Details of these properties for commonly used species are given in Appendix 'A'.

Arrises, more particularly those on exterior work, should always be eased or rounded. Where the arrises are sharp, the thickness of the paint or varnish film is reduced and it is at these points that breakdown of the finish often begins.

Specification

Drawings are most usually accompanied by a written description of both materials and workmanship required and appear either in the form of a 'specification' or are written into the bill of quantities. This written description, if it is to be of value, must be accurate, intelligible and, above all, possible to interpret in respect of materials and workmanship. For instance, to demand complete freedom from all defects, as frequently happens, is generally quite unnecessary and unrealistic. The quality specified should be linked with cost, obviously a higher quality of finish can be expected where final cost is relatively unrestricted, while on the other hand in certain situations 'defects' may be a design feature such as 'knotted' pine and rough textures.

It is also essential to differentiate between the materials and workmanship required for varying classes of work. BS 1186 'Quality of Timber and Workmanship in Joinery' (Part 1: Quality of timber and part 2: Quality of Workmanship) is an attempt to define 'minimum standards applicable to the production of joinery for the general requirements of housing and similar structures' and provide a more precise alternative to such time honoured expressions as "in a workmanlike manner" and "to the architect's satisfaction" which in practice depend entirely upon the judgement of an individual. This standard is obviously not set high enough for all classes of work but forms an excellent basis upon which appropriate clauses may be drafted. This British Standard specification contains a list of commonly available softwoods and many of the imported hardwoods considered suitable for replacement of softwoods grouped in uses other than decorative (for this see Appendix 'A'). See also Appendix 'B' and 'C' for other uses of timber and availability).

Moisture movement

Seasoned timber in service is subject to fluctuations of moisture content due primarily to changes in atmospheric humidity and such changes induce corresponding dimensional variations. The amount of movement varies with different species, e.g. Oregon pine tends to shrink considerably. It also varies with the direction of the grain. Movement along the grain is so small as to be negligible for all practical purposes but is considerable across the grain. This can be alleviated with some species by quarter sawing, e.g. oak and Douglas fir. Movement in all species is greater in the direction of the growth rings, i.e. in the tangential direction, being almost twice that in the opposite or 'radial' direction. The range is in the order of 1 mm per 300 mm (tangential) and .62 to .64 mm per 300 mm radially for each degree of change in moisture content.

This must be taken into account in the design stage and the necessary allowance made for movement within the humidity range in which the particular framework, boarding or panelling is to be placed.

It is essential in order to eliminate any large movement that the timber is dried to and fixed at a moisture content approximating that which will be met under service conditions. Notes on moisture content are given on page 67 of this book and details of the radial and tangential movements of timbers in common use may be obtained from FPRL Leaflet No. 47 'The Movement of Timbers'.

Work in the shop

It is not possible to lay down hard and fast rules as to the work to be executed in the shop and that which has to be performed on site. Each job must be taken on merit and in accordance with economic requirements together with the quantity of work involved. It is necessary at this juncture, however, to ascertain the precise meaning of 'work in the shop'.

Very little, if any, joinery or finishing work requiring machining, jointing, etc. has ever been performed 'on the site': even the smallest builder has traditionally done such work in his joinery shop and then conveyed it to the site for fixing. Today, however, there are few items within the category of finishings which are not readily available as mass-produced items from the large joinery factory. Special manufacturing of products, however desirable from the aesthetic viewpoint, does increase the cost probably more than the end product warrants unless a large volume of repetitive work is involved such as in the case of a large housing contract.

High class work inevitably demands special consideration and such work is usually supplied by specialist joinery works acting as sub-contractors who often not only supply the necessary items but also undertake the site fixing. This also applies to the larger contracting firms who commonly maintain their own joinery works.

Work in the shop consists, essentially, of drying or kilning the timber to the appropriate moisture content, accurately machining to the required section and finally assembling into components as necessary. Any preservative, moisture, fire retardant or other treatment specified are also performed in the proper sequence. Facilities for such work are only rarely, if ever, available at the site.

The assembled frames, etc. can, moreover, be stored at the works under proper cover and maintained in proper condition until required on the site for fixing.

Even where assembly is to be carried out on site, machining and pre-cutting can be economically done in the shop.

Work on site

Care and storage

Except perhaps on larger works, the facilities for the storage of joinery and particularly internal finishings which have been conditioned to allow moisture content, are rarely adequate and a carefully phased schedule of fabrication, delivery to site and fixing is necessary to reduce the period of site storage to an absolute minimum. Wherever possible delivery and fixing of internal finishings should be delayed until the heating is in use and the building properly dried out. These conditions are absolutely essential in the case of kiln dried hardwood flooring as explained under Moisture Content on page 18. Adequate ventilation in conjunction with the heating is necessary during the drying out stage in order to carry away the moist air.

External finishings and assemblies should be stored under cover, preferably in a dry, ventilated room, and stacked flat or in such a manner as to avoid deformation, twisting or accidental damage.

No timber which is to be incorporated in any part of a building should be left lying haphazardly on the site and fully exposed to the elements. Where it is anticipated, or known, that a period of site storage under adverse conditions may obtain, treatment with moisture retardants in the shop, particularly on the exposed end grain, is strongly recommended.

If polythene is used as covering provision has to be made for ventilation to avoid condensation.

Exterior work

The fixing of door and window frames should follow normal traditional practice. Fastenings should be close to the heads and sills but heads themselves should not be fastened. Generally the 'Design Guide' should be consulted for details.

Curtain walling

Curtain walling should be fixed back to the structure by bolted connections and never by screwing or nailing. Owing to the differing types of structure to which the walling may be attached, it is not possible to devise a universal method. Fixings should be flexible enough to admit of adjustment during erection, as well as subsequent moisture or thermal movements of both the walling and structure. They should receive special consideration at – or near – eaves, gables or corners where wind pressure and suction are most severe.

Panel walling

Panel walling which is fixed within the structural framework (as opposed to curtain walling which is fixed on the face) may sometimes be fixed in a similar manner to that described for window frames but it is recommended that bolted fixings be used in multi-storied buildings.

Wall sheathing

Wall sheathing is not regarded as normally necessary in the United Kingdom unless used for thermal insulation purposes or where it is required to develop a bracing or diaphragm action in conjunction with the supporting framework.

Wood sheathing should preferably be of tongued and grooved boards of a minimum thickness of 25 mm nominal where the cladding is to be fixed directly to it but in other cases the thickness may be in accordance with insulation or structural requirements. Horizontal sheathing is more commonly used but diagonal sheathing, applied at 45° angle, gives greater rigidity to the wall although the board wastage is more. Either type should be nailed with two nails at each support crossing and all vertical joints should be made over a support. The nails should be not less than 2½ times the thickness of the board.

Plywood sheathing should be of a minimum thickness of 8 mm when supported at 400 mm centres, 9.5 mm thickness at 500 mm centres and 12.5 mm at 600 mm centres. Nail sizes and spacing should be as indicated for plywood cladding. A special sheathing grade is available for this purpose. The use of a suitable vapour barrier, placed immediately behind the internal lining, is reccommended where plywood sheathing is used unless the construction is ventilated or a plywood interior finish is also used.

Particle board has been used as a sheathing but owing to the limited amount of experience it is felt that no general recommendations can be made here. Further details and instructions should be obtained from the appropriate manufacturer.

Wall cladding

Weatherboarding may be applied horizontally or vertically, and is generally available in various stock patterns or may be machined to any desired profile. Profiles and position of fixings of some of the most common forms are shown in figures 5 & 6. Nailing or screwing should be at the cèntre or at one edge of the board only to permit movement and avoid splitting. Concealed or secret fixing may successfully be employed, particularly with hardwood vertical boarding which is to be left untreated or finished with a clear varnish, but the boards should be not greater than 75 – 100 mm in width. Where some of the

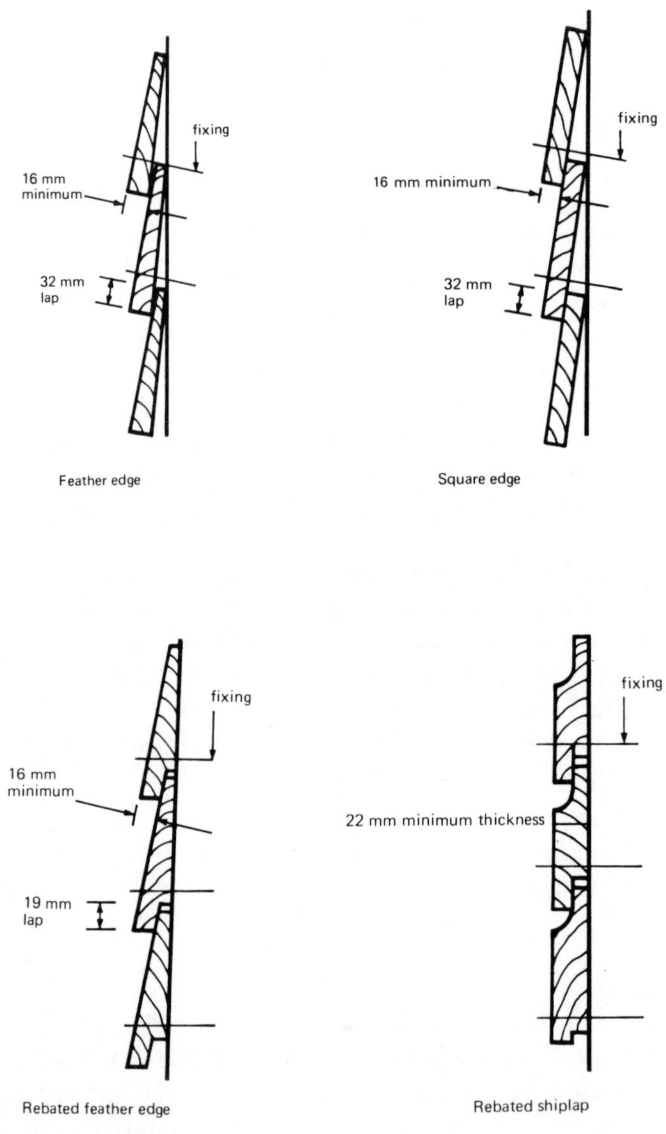

Figure 5. Types of horizontal boarding

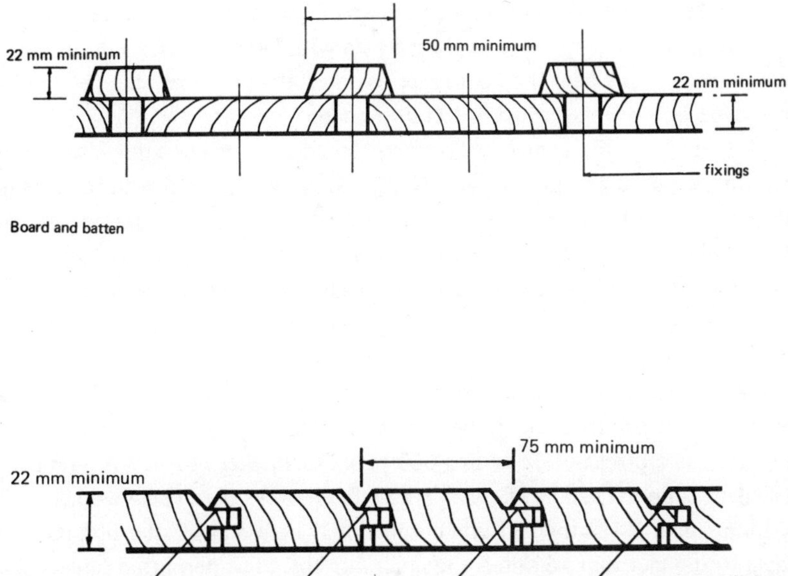

Figure 6. Types of vertical boarding

denser hardwoods are used pilot holes for the nails or screws are required and may also be necessary with most hardwoods when soft non-ferrous nails are used or the fixing is near the end of the board.

The bottom edge of all vertical boards and battens should be undercut to form a drip and the exposed end grain sealed with liberal applications of aluminium primer where the boards are to be painted, or clear sealer where the boards are to be varnished or left untreated.

Fixing nails should in all cases be rustproofed and consideration should be given to the use of aluminium nails. Nails cost little compared with the cost of the boards and labour of fixing and it is thus poor economy to use boards which will last for years then fasten them with nails which will rust badly and cause unsightly staining within a comparatively short period. The length of the nails should be 2 – 2½ times the thickness of the board or that part of the board through which it is fixed.

Where the boards are fixed directly to stud framing, battens or other supporting framework they should be of at least 25 mm nominal thickness and supported at centres not exceeding 600 mm. For thicker boards a pro-rata spacing of supports can be allowed. Where fixed over board or plywood sheathing, boards less than 25 mm nominal may be used but attention is drawn to the fact that 16 mm is the minimum thickness permitted under the Building Regulations. The Building Regulations also require that weatherboard clad walls must be at a minimum distance of 1 m*) from the nearest boundary.

For application to brick or block walls the boards should be fixed to 25 x 50 mm battens secured to the walling at centres as indicated above. Boards applied over wood sheathing of not less than 25 mm nominal thickness may be nailed direct but where applied over wood sheathing of a lesser nominal thickness, or over plywood sheathing, they should be nailed through the sheathing into supporting framework.

Building paper should be used as an immediate backing to the boards and it is important that this should not be of a highly vapour-resistive type since this may lead to condensation within the wall cavities. A number of types of kraft building paper fulfil this requirement: heavy building or roofing felts should never be used. The paper should be applied horizontally and lapped at least 100 mm both horizontally and vertically. Vertical joints must always occur over a stud in all cases except where suitable board sheathing is used.

Species of timber suitable for weatherboarding are given in Appendix 'A' and a guide to widths and lengths commonly obtainable is given in Appendices 'B' and 'C'.

See Fire resistance, Section 2.

Plywood cladding may be applied directly over framework as an alternative to weatherboarding but is more commonly used in inset panels. The plywood should be of exterior quality, preferably marine quality, in accordance with the recommendations contained in Sections 1 and 6.

Whilst plywood is relatively stable under moisture changes some small movement will take place and particular attention should be given to the sealing of all edges against moisture penetration. This applies more particularly where plywood is used in inset panels and 'beaded-in'.

Where sheets abut they should be set 1.5 – 3 mm apart, the inter-space filled with mastic and covered with a mould of timber or plywood at least 50 mm wide.

Where the plywood is fixed directly to the framework it should be supported along all edges and at intermediate positions as indicated below:

Intermediate support spacing (mm)	400	500	600
Plywood thickness (mm) face grain parallel to framing	9.5	12.5	12.5
Plywood thickness (mm) face grain at right angles to framing	9.5	9.5	9.5

As with weatherboard cladding all nails should be suitably rust resistant and the same remarks apply. The length of nails should be as follows:

Plywood thickness (mm)	6.5	8.0	9.5 and 12.5
Nail length (mm)	40	45	50

Brick or masonry cladding or facing is sometimes used with frame construction, for aesthetic reasons, and should be set away at least 25 mm from the face of the framing or sheathing. A building paper should be provided over the framework, or over the sheathing where provided, facing the brickwork or masonry. Rustproofed metal angle ties should be used to bond the facing to the timber framework or sheathing. These ties should be first screwed to the timber framing or sheathing at coursing levels approximately 800 mm apart horizontally and 400 mm apart vertically and subsequently embedded in the mortar joints. Where sheathing other than 25 mm nominal boarding is used it is wiser to screw the ties through the sheathing into the framework. Preference should be given to ties having a twist in the leg over the cavity in order that any rainwater penetrating the facing can be directed downwards and returned to the outside through weepholes at the base of the facing.

The thickness of the brick or masonry leaf should comply with byelaw requirements where these obtain or should be of the thickness normally required for the outer leaf in cavity wall construction.

Shingles can be used as cladding (see under roofs)

Yorkshire boarding can be used as cladding (see under roofs)

Roof sheathing

Boarding is commonly used as an underlay between the structural roof members and the final weatherproof covering. On roofs of low pitches the tendency for boards to curl in their width may cause irregularities in the surface of flexible coverings and allow the collection of pools of water. This tendency may be avoided by one or more of the following means:

a timber of the appropriate moisture content must be used (see under Moisture Content on page 18)

b the use of narrow width boards

c the boarding may be laid in the direction of the fall or, alternatively, diagonally across the supports

d the roof pitch may be increased

e the boards should be fixed as soon as the structural roof is complete and protected from weather exposure until the weatherproof covering is applied.

Where a flat roof is accessible for uses other than maintenance, and is covered with flexible material such as roofing felt, the boarding must be tongued and grooved and not less than 22 mm thick and span not more than 600 mm. The boards should be fixed with two nails at each support crossing with header joints made directly over a support and suitably staggered. The length of the nails should be not less than $2\frac{1}{2}$ times the thickness of the board.

The spacing of supports for other flat or sloping roofs up to 10 degrees pitch should be in accordance with the following table:

Minimum finished board thicknesses (mm)	16	19	22
Maximum spacing centre to centre of joists (mm) T & G	400	550	600
Maximum spacing centre to centre of joists (mm) Square edged	300–350	350–450	450–550

Plywood of a special sheathing grade can be utilised in place of boarding as a roof sheathing and has the additional merit of providing a much greater bracing effect.

For best results the sheet should be laid with the grain of the face plies at right angles to the supports with expansion joints of 1.5 – 3 mm between abutting sheets. Nails should be spaced at 150 mm centres along the edges and at 300 mm centres along intermediate supports and should be of lengths to suit the thickness of sheet as given under the plywood cladding. The eaves fascia should be in position before laying is commenced so that the bottom sheet can be nailed to the top edge of the fascia: the result being a stronger job and more watertight eaves.

Plywood for roof sheathing should be of a minimum thickness of 8 mm and the following board thicknesses and spacing of supports be used:

Nominal board thickness (mm)	8	9.5	12.5
Support spacing (mm)	400	600	800

The above thicknesses should be adequate where normal loads and very light surfacings are applied but where special loads are to be carried then calculations to determine the correct thickness of plywood and spacing of supports should be made.

The modern development of cross wall construction allows for the utilisation of plywood in stressed skin roofing units. These units may be utilised for both flat or low pitch roofs giving a light and rigid construction with an integral ceiling finish. See TRADA 'Design Guide'.

Particle board is also quite suitable for sheathing but it is not possible to generalise, and guidance on its use should be obtained from individual manufacturers.

Shingles may be described as wood tiles and form an excellent roof and wall covering. Their use should conform with local byelaws. The principal species of timber used are western red cedar, sweet chestnut and oak. Sawn western red cedar shingles are readily available in the United Kingdom but the other two species are somewhat limited in supply and are chiefly available in cleft form.

Cedar shingles are commonly imported in three lengths, 400, 450 and 600 mm with widths of from 75 mm upwards. The exposure depends on the pitch of the roof and should never exceed one third of the length of the shingle giving at least three thicknesses of shingle. By reducing the exposure on lower pitched roofs this is increased to four thicknesses. For roofs with a pitch of under 15 degrees cedar shingles should not be used. The minimum recommended pitch is 20 degrees. The maximum exposures (and batten gauge) for varying degrees of roof pitches should be as follows:

Shingle length (mm)	400	450	600
Maximum exposure (mm) 22 degrees or over	125	140	190
Under 22 degrees but not below 15 degrees	95	105	145

It is accepted practice in the United Kingdom to apply shingles on battens only to allow free circulation of the air underneath the shingles. The size and the battens vary with spacing of the rafters but 25 mm x 50 mm battens are adequate over the usual spacing of 600 mm. The following table gives a guide to rafter sizes and spans under the conditions stated.

Rafter size (mm)	50 x 75	50 x 100	50 x 125	50 x 150
Span (mm)	1400	2000	2500	3000

Design assumptions are for a total load (including 50 Kg/m^2 ceiling) of 150 Kg per square metre on plan as the slope is unknown, with a rafter spacing of 600 mm in each case.

Shingles should be laid with the first course at least double with the upper layer nailed so that the joints between the layers are not less than 38 mm apart. This is the minimum side lap and wherever possible should be increased. The shingles should not be tight butted but spaced about 6 mm apart to allow for movement. Joints in succeeding courses should be so placed that the joint in any one course should not be in a direct line with another joint unless this is at least three courses distant. Only two nails should be used for each shingle and should be so placed that the shingle in the next course above will cover the nail by at least 38 mm and with an edge distance of not more than 20 mm. Nails should be a minimum of 32 mm long for 400 and 450 mm shingles and a minimum of 38 mm long for 600 mm shingles. Rust resistant nails should always be used.

Shingles applied to walls present less of a weather resistance problem and a wider exposure may consequently be used. In single coursing the exposure should not exceed half the length of the shingle minus 12 mm. For 400 mm shingles, therefore the maximum exposure would be 188 mm: for 450 mm shingles 213 mm and for 600 mm shingles 288 mm. The same laying instructions as for roofs apply.

Close boarding or sheathing of a minimum nominal thickness of 25 mm may be used on wall surfaces and the shingles nailed direct over building paper to the sheathing. In all other cases battens, firmly secured to the framing, brick or block walls, should be used.

Yorkshire boarding is often used for farm storage buildings and laid as shown in figure 7, it can also be used as a cladding.

The boards are usually 19 – 25 mm in thickness by 150 – 175 mm wide, and should have a 10 mm diameter groove about 10 mm from the edge on the upper surface and a slight rebate on the underside edge. Each board should should be spaced about 3 mm from the adjacent board and fixed by one 100 mm rustproofed wire nail through the centre of the board into the purlin. This centre fixing allows for movement of the board and avoids consequent splitting. The boards should be separated from the purlin by means of rustproofed hobnails, previously driven into the purlin, with their heads projecting at least 6 mm.

Figure 7. Yorkshire boarding

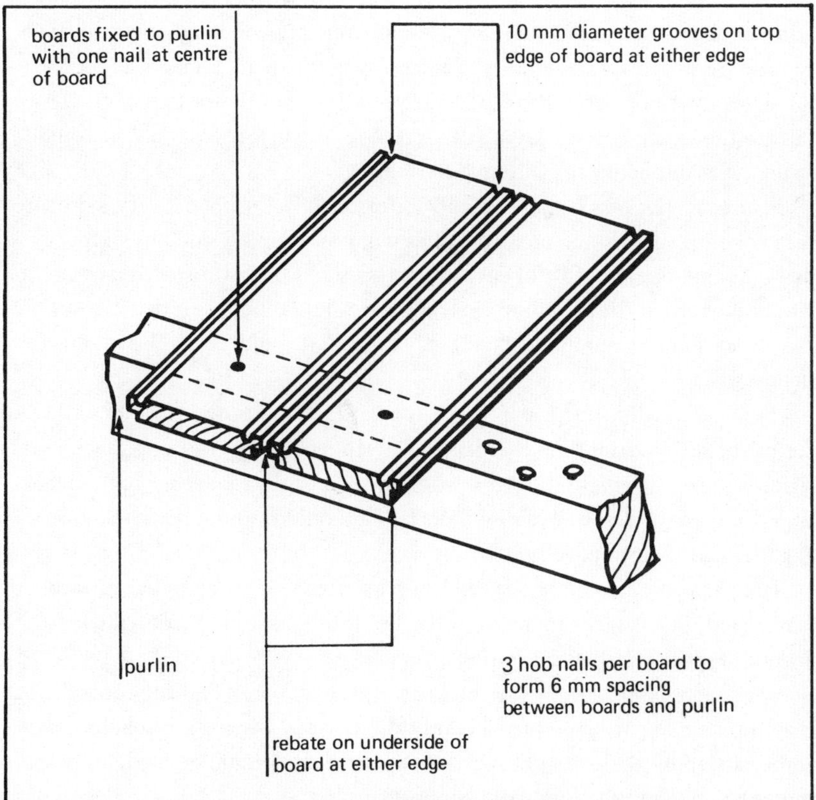

Most home-grown timbers may be used for this purpose as well as such imported timbers as whitewood, redwood, western hemlock and Douglas fir.

Only properly seasoned timber should be used which should be preservative treated preferably by impregnation under pressure.

Durability

An important consideration in the selection of boarding for cladding is durability and stability, i.e. moisture movement. In the TRADA Memoranda dealing with the Building Regulations sheet 7.1 deals with methods of treatment and makes some recommendations. It further classifies timbers suitable for external cladding in groups of durability. This sheet is reproduced in Appendix 'G'.

Interior work

Sub-flooring

Sub-flooring of softwood and hardwood boards, plywood and, more recently, particle board are commonly used as a base for such materials as cork, linoleum, thermoplastic tiles etc., which require a continuous support. Details of the treatment and laying of such covering materials should be sought from the manufacturers of these coverings.

Board sub-flooring is more usually required only to support very light surface coverings and, since such coverings will not normally add any extra rigidity, the boards should be of such thicknesses and laid over such spans as would be normal for softwood flooring. Care should be taken that the surface is level and it should be examined to see that there are no protruding nail heads.

Plywood sub-flooring is employed for the same purpose as that described for board sub-flooring above. It has the advantage of contributing much extra rigidity to the floor structure as a whole whilst, at the same time, offering a smooth, firm surface free of any defects that could result in depression or fracture of the floor covering.

The plywood may be of sheathing grade except where a sanded surface is required. Bearers, cut in between the floor joists, should be provided to ensure support for all edges in the latter case.

For general applications the thickness of the plywood and support spacings may be as under. Where unusual conditions exist or when abnormal loads are to be carried by the plywood sub-floor calculations should be made to determine the plywood thickness and support spacing.

Plywood thickness (mm)	12.5	16	19*
Support spacing (mm)	400	500	600

*May be 12.5 mm where 25 mm nominal hardwood strip flooring is used and laid at right angles to supports.

For best results the sheets should be laid with the grain of the face plies at right angles to the supports with expansion joints of 1.5 – 3 mm between abutting sheets. Nails should be 60 mm long for 12.5 mm and 16 mm thicknesses and 63 mm long for 19 mm thickness and spaced at 150 mm centres along the edges and at 300 mm centres along intermediate supports. If screws are used for fixing they should be spaced at the same centres as for nails and should have a length of at least 12.5 mm greater than the thickness of the plywood.

Plywood is sometimes used as an intermediate layer between an existing board floor and a parquet or similar type overlay. The thickness of the plywood should be at least 3.2 mm where laid over 25 mm nominal softwood boarding. It is recommended that the plywood be laid diagonally to the run of the existing boards which should be level and free from protruding nail heads, etc.

Particle board can provide an excellent sub-flooring when correctly laid. Advice on suitable types, thicknesses and spacing of supports should be obtained from respective manufacturers.

Flooring

The continued popularity of wood as a flooring material arises from its outstanding natural properties among which are durability, resilience, thermal insulation and attractive appearance. It is emphasised, however, that technical knowledge and skill must be applied to both the laying and the sub-floor construction if the best service is to be ensured. The flooring in a building receives much harder treatment than most other finishes and deserves very careful consideration and selection. This applies more particularly to the various types of hardwood flooring.

Softwood flooring is more generally fixed by the building contractor but it is, nevertheless, essential that timber of the correct moisture content be used and that precautions are taken against excessive movement. The use of narrow width boards will help to minimise the effects of any shrinkage.

In the past it has been the usual practice to fully cover softwood flooring with carpeting, linoleum, etc. but this extra expense can be spared and an attractive surface obtained given care in the selection and laying of the boards together with proper application and maintenance of a modern durable floor seal.

Hardwood flooring whether in strip, block or parquet form requires expert laying and should always be undertaken by a specialist flooring contractor. It must be remembered that the choice of a hardwood flooring is not only dependent upon the choice of species for the desired colour and decorative effect but also on the amount of traffic which the floor is to receive.

Detailed information on the various types of flooring and species suitable for different purposes is given in TRADA booklet 'Wood Flooring'.

Plywood flooring is used mainly in the form of parquet tiles for domestic premises for surrounds to carpets in living and bedrooms and corridors where foot traffic is relatively light. Special tiles with highly decorative hardwood veneers are available from specialist manufacturers who should be consulted with regard to availability and method of fixing, etc. The tiles are most usually laid over a timber or similar sub-floor and glued and pinned.

Particle board flooring. Properly laid and finished particle board can provide an excellent floor but suitability of any particular board for the purpose should be ascertained from the manufacturer.

Wall and ceiling linings

Wall and ceiling linings of timber boards, plywood or particle board may be used either as a complete room finish or as a foil to other finishes in accordance with aesthetic requirements. Such linings are easily applied, and, when used in conjunction with frame construction, greatly facilitate the installation of services which may be run within the wall thickness thus eliminating the expensive process of chasing usually required with conventional construction. Condensation, plaster cracking together with the pervading atmosphere of cold and dampness found in traditionally built houses is avoided.

Attention, however, is drawn to Building Regulations requirements in respect of internal resistance to fire of certain walls and ceilings. See Section 2 and Building Regulations.

Board linings sometimes called plank, may be used either vertically or horizontally and the application is greatly facilitated by the use of a tongue and groove system. As an alternative the boards may be laid in shiplap or other patterns to obtain the desired effect. The general application of all types is similar to that for exterior weatherboard cladding particularly with reference to pilot holes for nails or screws where hardwoods are used, and length of nails to be used.

For general applications as a lining to both walls and ceilings the minimum board thickness and spacing of supports may be as under:

Nominal board thickness (mm)	12.5	19	25
Support spacing (mm)	600	750	900

In vertical applications the boards will generally run the full height required but in horizontal applications some heading joints may be necessary. In either case, where a heading joint is to be made, the supporting member should be at least 50 mm on face.

Timber may be regarded as having greatei resistance to impact than most other lining materials. This resistance is dependent mainly upon the density of the species used and, with some exceptions, hardwoods will have greater resistance than softwoods. Where this property is required the spacing of the supports given in the table above should be halved.

For application to brick or block walls the boards should be fixed to 25 x 50 mm battens, plumbed and secured to the wall at the centres indicated in the table above.

Lightweight fittings may be secured to boarding of the thicknesses indicated above but heavy fittings will require adequate supports or bearers behind the boards to which fixing should be made. A number of proprietary board lining systems are at present available and fixing should be in accordance with the manufacturer's or supplier's instructions.

Reference should be made to Appendices 'B' and 'C' for a guide to the widths and lengths in which various species may be obtained.

Plywood linings are easily installed and can produce considerable saving in both labour and material if rooms are planned to use standard sheet sizes. A number of types of specially surfaced or textured plywood are obtainable where special effects are desired. Generally plywood should be specified and used in accordance with the recommendations contained in Sections 1 and 7. Appendix 'E' gives grades, plywood species and sizes commonly available.

For general application as a lining to both walls and ceilings plywood should be supported on all edges and at intermediate positions as indicated below:

Minimum plywood thickness (mm)	6.5*	9.5	12.5
Maximum intermediate support spacing (mm)	400	600	900

*may be applied over 600 mm centres of support spacings if additional support is given at centre of length at right angles to intermediate supports or if the face grain is at right angles to intermediate supports.

Length and spacing of nails should be as given for plywood cladding. For application over existing brick or block walls or ceilings the plywood sheets should be fixed to 25 x 50 mm battens secured to the existing construction at positions indicated above. Care should be taken, where fixing to a ceiling is contemplated, to ascertain that the existing construction can safely carry the additonal loading. Lightweight fittings may be secured to plywood linings of 9.5 mm thickness or over but heavy fittings will require additional supports or bearers behind the plywood to which fixing should be made. Several proprietary plywood lining systems are at present available and fixing should be in accordance with the manufacturer's or supplier's instructions.

Particle board is now being increasingly used as a wall lining material and a number of types with specially veneered or decorative surfaces are now obtainable. Particle board should be specified and used in accordance with the recommendations contained in Sections 1 and 9.

It is not possible to generalise on details of fixing etc., and such data should be obtained from individual manufacturers.

Section four: preservation and finishes

Preservation

Timbers and plywoods of different species vary considerably in their durability. Whilst it is advantageous to use timbers with high natural durability, it may not always be economic. Timbers with low durability may be used, providing they have been treated with a preservative if they are to be used in situations conducive to decay. These situations are considered to be as follows:

a In contact with the ground
b wholly enclosed in brickwork, concrete or masonry
c in any situation in which adequate ventilation cannot be provided
d in any situation in which equilibrium moisture content of the timber is likely to exceed 20 per cent
e timber classified as of low durability, or with a fair amount of sapwood, and used in circumstances in which there is reasonable doubt of service conditions in regard to temperatures, ventilation and humidity
f in special cases in areas where fungal and insect attack is known to be prevalent
g where subject to attack by marine or other water-borne organisms.

Timbers and plywoods classed as durable and very durable are frequently resistant to preservative treatment, and as such may be used in any of those situations without such treatment, but it is considered a wise precaution to apply preservatives if any sapwood is present.

Types of preservatives and characteristics

Wood preservatives fall into three main types:
i Tar oils
ii Water borne
iii Organic solvents.

Each of these have their advantages and all three are effective against decay and wood destroying insects.

Tar oil preservatives

i they are very resistant to leaching and so are particularly suitable for treating timbers to be used for exterior work, in water, or buried
ii they are not, as a rule, corrosive to metals
iii they are difficult to paint over
iv they tend to 'creep' in plaster or other absorbent materials with which they may come in contact
v they have a strong odour which, under certain conditions, may be picked up by foodstuffs which are near to, although not in actual contact with, treated wood
vi freshly treated timber is more flammable than untreated, but after a few months the volatile fractions of the oil evaporate and the fire hazard is greatly reduced. There is evidence that creosoted timber is then no more flammable than untreated wood.

Water borne preservatives

i they are less viscous than tar oils (creosote), giving good penetration
ii they are usually odourless
iii they may be painted over when the treated wood has dried
iv they are usually non-creeping and do not stain
v they are non-flammable
vi they are cheaper and easier to transport than other types of preservatives, as they can be shipped in powder form
vii they are non-leaching and suitable for internal and external use but there may be a tendency for some of the older types of water-borne preservatives to leach out of the wood when used in contact with the ground or in water
viii re-drying of the treated timber is necessary
ix certain of the water-borne preservatives may be used for treating cases for the conveyance of foodstuffs without risk of contamination or taint.

Organic solvent preservatives

i they are resistant to leaching, are reasonably permanent, and so are suitable for both interior and exterior use
ii they may be painted over after the solvent has evaporated
iii they are usually non-creeping and do not stain
iv they are not corrosive to metals

v when dissolved in volatile solvents they usually penetrate the wood rather better than other types of preservative and so are more suitable for brush application, spraying or cold dipping
vi volatile solvents may be somewhat flammable; care is therefore necessary when using them in confined spaces and in storage. After the solvent has evaporated, treated wood is not more flammable than untreated wood
vii some have a strong odour which may be picked up by certain foodstuffs, even if not in actual contact with the wood; others are free from this odour
viii with a suitable solvent, swelling of the wood does not occur and so they can be used on wooden parts already cut to accurate sizes
ix some are suitable for treating horticultural and agricultural timbers as they are non-injurious to plant life after the solvent has evaporated.

Preparation of the wood

Adequate preservation essentially requires not only a good preservative but also a satisfactory method of application. Further, the timber must also be in a suitable condition to absorb the preservative. Timber should be seasoned before treatment. The optimum moisture content to which timber should be seasoned varies with the species, the size of the timber and also the type of preservative used. When water-borne preservatives are used, the moisture content may be somewhat above the fibre saturation point, i.e., about 30 per cent. With organic solvent type preservatives the moisture content of the timber should be below 22 per cent; in the case of creosote, 27 per cent. Where round timbers are used all bark and inner bark should be removed in addition to the timber being air dry.

Any cutting, shaping and boring should be carried out before treatment. If this is impossible a liberal application of the preservative should be made to all cut surfaces.

Methods of application

These are: non-pressure and pressure

Preservative protection, which is the most important criterion, can be achieved by both methods provided proper care is exercised in selecting the method of application appropriate to the type of preservative and to the end use of timber. More detailed information on this subject will be found in:

a BWPA TRADA booklet, 'Timber Preservation'.

b BS 1282: 'Classification of Wood Preservatives and their methods of Application'.

c Code of Practice CP 98 'Preservative Treatments for Timber used in Buildings'.

Non-pressure methods are by brushing, spraying, dipping, steeping or hot and cold open tank, the last three being generally more effective than the first two since deeper penetration of the preservative can be achieved. Organic solvent type preservatives are generally applied by non-pressure methods and owing to their superior penetrative qualities give satisfactory results. With certain water-borne preservatives such as boron compounds, after the green or unseasoned timber has been dipped or passed through a spray tunnel the timber is stored under retarded drying conditions for a period of 2 – 6 weeks, so as to permit diffusion of the preservative throughout the wood. This method is not normally used for timbers where there is a severe leaching hazard. Coal tar oil types other than creosote are normally applied by non-pressure methods.

Of the pressure methods there are two types, full cell and empty cell. In the full cell treatments the cavities of the elements or cells which go to comprise wood are filled with preservatives; in the empty cell only the walls remain coated with preservatives after treatment. The main advantage of the empty cell treatment is with tar oils where there is less bleeding out of the preservative and 'cleaner' timber is achieved.

Creosote is normally applied by pressure impregnation as are water-borne preservatives of the copper/chrome, copper/chrome/arsenic and fluor/chrome /arsenate/dinitrophenol types. A certain amount of treatment is also carried out by the hot and cold open tank method and some creosote is applied by dipping and brushing. For cladding, these days mainly the non-organic solvent types are used or copper/chrome/arsenic.

Effective life and renovation

Timber and plywood which have been effectively treated with a preservative last for long periods.

Where these treated materials (particularly by brush and spray) are likely to be subjected to abrasion or mechanical wear that may expose untreated surfaces, or where they are likely to be subjected to weathering, they will require periodic inspection which may involve re-treatment.

Specification

It is recommended that the appropriate British Standards (see Bibliography) is invoked for the pressurised treatments and that where a proprietary product is named the makers' instructions are followed. See Appendix 'G'.

Fire retardant treatment

As stated in Section 2 in many cases timber has to comply with Class 1 surface spread of flame requirements under the Building Regulations.

Specification

Compliance with Class 1 surface spread of flame requirements usually needs to be specified. For impregnation treatments this requires a retention of salts in the timber of 40 kg/m^3.

Types and characteristics

There are two types of fire retardant treatment for timber. The first type is a water solution of various inorganic salts which has to be forced into the wood under a process of vacuum and pressure impregnation to be effective. The second type is a surface coating which may be applied in much the same way as paint.

Impregnation with salts is potentially more effective but since the achievement of maximum effects may be very expensive, it is usual to employ a process which gives a product complying with Class 1 spread of flame for both impregnation and surface coating. There is therefore nothing to choose between the two types as far as test requirements are concerned.

Impregnation gives timber a permanent treatment on all surfaces (except under conditions of leaching where the salts wash off the surface layers). The timber has to be cut to final size before treatment and it usually needs to be dried after treatment. Under high humidity conditions it may become wet and the salts cause difficulty in decorating and glueing. Preservatives are normally incorporated with fire retardants.

Surface treatments can be applied to timbers which are already in position but it may be difficult to treat all surfaces. They require renewal at intervals. They possess little resistance to damp but are available in a range of colours, resembling other matt decorative paints.

Appendix 'H' gives performance data on test results for timber, plywood and various timber products and treatments to achieve Class 1 performance.

Preparation of the wood

For impregnation with fire retardants the timber should be below 18 per cent moisture content and cut to final size. Before treatment with surface fire retardants previous finishes must first be removed.

Application

Impregnation fire retardants must be applied by a specialist firm in a vacuum and pressure cylinder. Surface coatings may be applied by brush, spray, dip or any other standard method.

Effective life and renovation

Impregnation treatments last for ever unless they are washed out by continuous leaching. Renovation would involve dismantling the structure completely.

Surface coatings possess the normal life of a decorative coating and require renewal when the surface has become dirty and chipped.

Paints – types and characteristics

It is traditional to apply paint to wood as a three or four coat system. The first coat or primer serves to seal the surface and provide a basis for subsequent coats. The next coat, undercoat, contributes to the thickness of the paint film and starts the build up of colour in the film. For exterior purposes or high quality interior application, two coats of undercoat may be applied. The final or finishing coat gives the film its resistance to weathering, abrasion etc. and determines the appearance of the coating.

Primers – the standard lead based oil primer has been largely superseded by less costly and less toxic materials but is still the basis for comparison of the performance of other types.

Leadless primers vary widely in price and in performance from products almost equivalent to lead primers to those with very low durability and weathering resistance. Acrylic primers are being increasingly used for both interior and exterior applications. They have the advantage of being water thinned, non-toxic and convenient to use but, when exposed to the weather, allow more moisture penetration than other types. Aluminium wood primer is recommended where the sealing back of resin or oily preservatives is necessary. This type is also very effective as a moisture barrier and particularly useful for the sealing of end-grain surfaces.

Oil based undercoats and finishing coats have given way almost entirely to paints based on synthetic resins. Virtually all the solvent based exterior paints currently used are of the alkyd type and of a high gloss finish to reduce the retention of dirt. The same paints in gloss and matt or eggshell finishes are available for interior use.

The other major class of paints for interior use are the emulsion paints based on vinyl and acrylic resins. They have the advantage of convenience of use, rapid drying and relative freedom from odour. Both these and the alkyd paints are produced in thixotropic form to enable the application of thick coats without sacrificing surface finish.

Water thinned acrylic paints which give a gloss finish have already become available for exterior use. The inability to provide a gloss surface held up the exterior use of this class of finish for several years but now that this problem has been overcome, an alternative to the alkyd exterior paints is available. There is as yet little evidence as to their effective life but it should be at least equivalent to that of the more conventional finishes.

For interior use there are many specialised finishes including lacquers, anti-corrosion paints, fungicidal paints and more recently polyurethane paints with the advantage of high impact, abrasion and stain resistance. A further specialist group of paint finishes are those combining the decorative function of a paint with the protective function of a fire retardant coating. See Appendix 'H'. These are available in a variety of colours and surface textures but most are suitable for interior use only.

Textured finishes based on a variety of building media are available and are particularly useful for the treatment of exterior plywood. A variety of textures are available and a service life considerably longer than that of conventional paints is claimed for these products.

Maintenance and renovation

Painting over a previously painted surface is possible provided the old paint film is not too thick, if it is the same type of paint, the previous surface is intact, and the surface is clean and, preferably, rubbed down with abrasive. Where previous paint is failing or it is too thick it must be removed by the use of a paint stripper or blow lamp and a scraper.

Preparation of the wood

Before painting, wood surfaces must be dry (below 18 per cent moisture content) and clean.

If a flat surface is required the wood must be fine sanded and with coarse grained woods, the grain must be filled with a wood filler which is applied by rag and rubbed off across the grain when dry.

Application

Brushing is still the common procedure in buildings, though roller application is particularly suitable with flat surfaces and with emulsion paints, and spraying, dipping, flow coating, electrostatic coating etc. are usual in factory applications. When attempting to force-dry coatings on timber the likely shrinkage due to moisture content changes must be considered as it may cause cracking of the finish.

Stains – types and characteristics

There are three types of stains, water stains, spirit stains and oil stains. Each may be supplied ready made up or as a solid powder to be dissolved as instructed in the appropriate solvent.

Water stains are made up in water. Spirit stains are made up in industrial methylated spirits. Oil stains are made up in linseed oil with white spirit thinners and they may contain insoluble pigments.

For hardwoods/softwoods/'difficult' timbers

The main difference between the staining of hardwoods and softwoods is that most hardwoods are less permeable to liquids than most softwoods. Softwoods usually absorb stain more readily and more will be absorbed by the soft springwood bands than by the hard summerwood, thus emphasising the grain. More dilute stains are needed to give a colour on softwoods than on hardwoods. Hardwoods with large vessels or coarse grain such as oak, elm, ash, will absorb more stain in the vessel structure so showing it up. Stain can usually be applied to hardwoods liberally and fairly concentrated.

Oily timbers such as teak or resinous timbers such as pitch pine do not stain readily with oil stains. They may be washed with detergent or white spirit, or water or spirit stains may be used.

Preparation of the wood

The surface of the wood should be cleaned and smoothed down with fine abrasive applied along the grain. Blemishes should be filled with plastic wood, plaster of Paris or putty, except where water stains are used when stopping can be done after staining.

Water stains tend to swell the surface fibres of the wood and raise the grain. This can be largely avoided by pretreating the wood with water, allowing to dry, and rubbing down the raised grain with abrasive before application of the stain.

If light staining is required on softwoods it may be necessary to pretreat the wood with hot weak size, or thinned shellac knotting to reduce its absorption.

Application

Stains may be applied by a brush, rag or spray. Excess should be wiped off with a rag. Care should be taken, by initial thinning of the stain, to avoid applying too much, particularly with softwoods. Spirit stains require particular care because they dry so quickly.

Effective life and renovation

Stains are usually protected by a clear resinous finish and provided the finish is preserved intact the stain will not need renewal. If the finish is damaged it should preferably be removed with a stripping solvent since this does not damage the stained surface. If the stained surface is damaged it may be possible to patch it with stain but usually the whole surface will require to be sanded down to untreated wood and the process repeated.

Natural and clear finishes - types and characteristics

With both interior and exterior use of timber it is often desired to retain the natural appearance of the wood for decorative purposes. There is a wide variety of natural finishing materials available. These products vary in their suitability depending on their conditions of exposure. The requirements of a finish for interior panelling differ from those needed to maintain an acceptable appearance on external cladding.

Broadly speaking, natural finishes may be considered in two groups, those which penetrate the wood surface and those which remain on the surface to give a tough resinous film.

For further properties and performance data of external finishes. See TRADA publication 'Maintaining Timber Exposed to Weather'.

Theoretically the preferred exterior treatment for timber would be a clear resinous film which would protect the wood from colour loss while retaining or enhancing its decorative properties. Unfortunately no such material is known which does not require regular maintenance every two to three years.

An alternative treatment is available in a range of materials grouped together as water repellant preservative stains. They may be 'natural' (designed to enhance the original or weathered appearance of the wood) or pigmented in some way to produce special effects.

So there are two distinct categories of treatment.

1 Resinous clear finishes (varnishes, natural and synthetic)
2 Water repellent preservative stains.

The differences between these two groups are considerable.

Group 1 provides in theory the most effective means of maintaining the original appearance of timber by creating a protective transparent film over the surface. In addition to protecting it from wetting it screens out most of the damaging ultra-violet light responsible for the colour fading. But, as the coating materials are themselves subject to deterioration they do not – in their present state of development – represent a complete practical solution for 'natural' finishes. Durability can be greatly increased by the application of one or preferably two coats additionally to what is normally specified. All varnishes tend to dry slowly and to be soft, but they possess good elasticity and give thick protective coats. Once varnishes fail they will do so by cracking and considerable labour is usually required for preparing the surface for retreatment. Also varnishes are liable to deteriorate more rapidly on sharp edges and arrisses.

Mention should be made here of Lacquers. A lacquer is defined as a solution of a natural or synthetic resin in a suitable solvent. They therefore tend to dry quickly and give thin coats. Resins which may be applied by means of solvents include shellac, nitro-cellulose, polyvinyl acetate and urea formaldehyde.

Preparation of wood for varnishing

Before application of a clear finish the wood must be sanded down clean and smooth. All traces of wax, alkali and previous finish must be removed. If a smooth glossy final surface is required it will be necessary to fill the grain of coarse timber. A sealer varnish should first be applied and then a good quality wood filler of suitable tint applied by rag and wiped off when dry across the grain.

Staining (see page 58) may be done before filling if required.

Application

Application of clear finishes may be made by brush, rag or spray.
Special dual feed sprays or curtain coating machines are required for factory application of two-can plastics.

Effective life and renovation

The frequency of renewal of a clear finish varies widely with the situation and the finish used. French polish on an exterior surface will require monthly renewal. A resinous finish on an ornamental sideboard may never need to be renewed.

The life of clear finishes on exterior timber surfaces is much less than many seem to expect. Four coats of a good quality exterior grade varnish have a life about three years on exterior cladding. Renewal must be attempted before the coating has failed. The old coating is cleaned down thoroughly and preferably treated with light abrasive action before applying two further coats of varnish. If the varnish has been permitted to fail extensively the colour will be washed out of the wood and it may be necessary to apply a light stain before revarnishing.

There are no clear finishes which have a longer life on exterior timber than a good quality exterior varnish. If maintenance at three year intervals cannot be carried out some protection may be given to the wood by a water-repellent stain which does not prevent loss of colour or surface roughening of the wood but tends to replace the lost colour and retards surface breakdown.

Internally the life of resinous finishes (varnishes, lacquers, two-can plastics) depends on the amount of wear they get. Wax and oil finishes should be restricted to surfaces not subject to wear. Where hard wear or resistance to water or stains is required the two-can type of finish are available. Renovation depends on the type of finish. It is usually most satisfactory to remove the old finish by a suitable solvent or by abrasion. Where wear is not too great it may be adequate to clean off, and treat lightly with abrasive before re-treating. In general the same type (or even brand) of finish should be used for renovation. Special difficulties may arise through mixing them up. Wax finishes must be removed before applying resinous finishes as they inhibit drying and reduce adhesion.

Effects on solid wood or veneers

Transparent finishes darken the wood. Finishes that are not completely transparent are usually yellowish or brownish and these darken the wood even more than transparent finishes. Of all finishes the polyvinyl acetate latex type change the colour of timber least.

All resinous finishes seal the pores of the wood and so prevent it from holding dirt and reduce its ease of staining. They also protect the surface from wear and change of colour by leaching.

For hardwoods/softwoods/'difficult' timbers

There is no essential difference between clear finishes for hardwoods and softwoods. For coarse plywood, where there is excessive absorption of finish, a special primer may be necessary. Shellac is suitable for internal work but a gold size type varnish is probably most suitable externally.

Oily (e.g. teak) or resinous timbers (e.g. keruing, yang, pitch pine) may give difficulty. Removal of the oil and resin by washing with soap and water or white spirit solvent followed by priming with a sealer such as shellac or gold size usually avoids trouble. Inhibition of setting of varnish or lacquer occurs with certain timbers. Oil varnishes are only affected, seriously, by a few species such as the *Pterocarpus* and *Dalbergia* but polyester resins fail to set satisfactorily on many species. A primer of polyurethane lacquer avoids this difficulty.

Group 2. Although the composition varies, all these materials have the characteristics in common of penetrating to a varying extent the surface of the timber. The water repellent action mainly consists of preventing the wetting of the wood surface by interposing a water-repellent layer rather than by relying on a low permeability film.

They are mostly based on the 'Madison Formula' developed by the Department of Agriculture and Forest Products Laboratory of the U.S.A.

The original Madison Formula on a percentage of weight basis

	Cedar	Light Redwood	Dark Redwood
Paraffin Wax	2.5	2.5	2.6
Zinc stearate	0.3	0.3	0.3
Turpentine or Paint Thinner	16.3	15.9	18.4
Pentachlorophenol Concentrate	12.0	11.9	12.2
Boiled Linseed Oil	57.6	56.2	58.1
Burned Sienna in Oil	6.9	13.4	2.3
Raw Umber in Oil	4.4		1.5
Indian Red Oxide in Oil	–	–	4.6
	100 %	100 %	100 %

These materials as can be seen have a high penetration value and a low pigment content and so the natural grain pattern of the wood is not obscured.

They are not so susceptible to sharp edges as varnishes and breakdown occurs mostly by erosion and leaching.

Preparation of surface

Only washing and wiping down is required and the removal of surface dirt. The treatment is considerably less costly than for varnishing.

Application

Ideally this should be by dipping and especially the **end grain** should be thoroughly treated, but in practice this is rarely possible and a minimum of two coats but preferably four should be applied by brushing including any exposed end grain.

Section five: specification of timber

Selection

With the great variety of timbers, it is important to know in what form they are available and in what sizes. This general data is presented in Appendices 'B' and 'C'

In order to avoid confusion the standard and botanical name of the timber selected should always be quoted; for this information British Standard 882 and 589: 1955 'Nomenclature of Commercial Timber' should be consulted. Also BS 1186 Part 1.

Colour will vary and reference should be made to Appendix 'A' for this information. Colour can be altered by bleaching or staining which however is not always uniform. Distortion in the grain may give rise to unsatisfactory behaviour in certain conditions of use such as large window sections and curtain walling. Material for these purposes should be carefully selected for freedom from irregular grain. The above comments are particularly applicable to hardwoods and for high class work.

Specification

In specifying timbers the following should always be stated:

1 Standard and botanical names

2 Exact purpose for which the timber is to be used – to enable the supplier to select suitable material

3 Situation in which the timber is to be used. State moisture content and specific conditions (see table and also remarks under flooring below. Refer to BS 1186 Part 1. 1952 and amendments.

4 Whether preservative treatment is required.

5 Any British Standard or Code of Practice appropriate to the above.

Limitations

The sizes of timbers are limited to the dimensions of the trees from which they have been converted. If a timber is classed as being only available in 'shorts' or 'blocks' it is useless to specify this wood for joinery purposes. Rhodesian teak, loliondo, muhuhu come under this category. For details of sizes of softwoods and hardwoods see Appendices 'B' and 'C'.

Certain timbers are known to stain badly in contact with iron– e.g. oak, sweet chestnut, afrormosia, western red cedar, with such timbers non-ferrous fastenings should be used.

Gum exudations occur, in gurjun/keruing/yang, guarea to a certain extent, and occasionally, agba. Washing with a solvent will help to minimise this. Oak is liable to discoloration in contact with cement.

Fugitive colours, as in mansonia and purpleheart and iroko, can also occur. Iroko will change from the bright yellowish brown to a uniform deep brown colour and even opepe will, during the passage of time, darken very considerably. Timbers with fugitive colour characteristics are given in Appendix 'A'.

Moisture content

The importance of moisture content cannot be stressed too highly. More trouble is caused through incorrect specification of moisture content, or failure to specify it, than any other way.

Wood is a hygroscopic substance which always reacts to changes in atmospheric conditions, but if it is dried to a moisture content in equilibrium with its surroundings slight seasonal variations will not give rise to movement of the material which can cause warping and shrinkage.

Another factor to be taken into consideration is that heating methods used today and the improved insulation of buildings are bringing the equilibrium moisture content of timber to a lower level than was the case previously. Thus for centrally heated buildings the moisture content of timber should not exceed 12 per cent.

For structural members and carcassing timber a moisture content of up to 20 per cent is acceptable. For timber which is to be painted or varnished moisture content should not exceed 18 per cent.

The table on the next page will give a general indication as to moisture contents for various uses.

TABLE OF MOISTURE CONTENT OF TIMBER FOR VARIOUS PURPOSES

Average moisture content attained in use in a dried out building (%)

EXTERNAL	
Doors	12 – 18
Windows	12 – 18
Curtain walling	17 – 20
Panel walling	17 – 20
Weather boarding	17 – 20
Wall sheathing	17 – 20
Shingles	17 – 20
Roof sheathing	17 – 20

All timber used in external situations on a building should be fully air dry, i.e. 18% moisture content at the time of installation.

INTERNAL	
FLOORING	
Intermittent heating	12 – 15
Continuous heating	9 – 11
Under floor heating	6 – 10
WALL & CEILING LINING	
Intermittent heating	12 – 15
Continuous heating	10 – 12
DOORS	
Intermittent heating	12 – 15
Continuous heating	10 – 12
JOINERY	
Intermittent heating	12 – 14
Continuous heating	10 – 12

Flooring timbers are in a different category since much of this work is subcontracted in which case the subcontractor should be fully informed as to

a method of heating, i.e. underfloor, ceiling or other forms of central heating.

b maximum temperature at screed and floor level,

c type, nature and depth of screed (if any) or other forms of sub-floor or floor structure,

d if a screed forms the under-floor when it was laid and if adequately protected against rising moisture and its general characteristics (preferably a working drawing).

In any case all wet work should be completed and the building should have an ample chance to dry out prior to laying the floor. Failure to observe these requirements will increase the moisture contents beyond the required limit.

In case of intermittent heating it may not always be possible to kiln dry a limited quantity of wood to a slightly higher moisture content than that required for continuous heating and slight shrinkage may occur. If normal kiln dried timber is used as for continuous heating slight expansion would occur when the heating is not in operation and allowance for this in the workmanship of the actual laying should be made by the provision of expansion joints or in the margins.

This latter method is probably preferable.

In the case of underfloor heating which is becoming less common these days the moisture content should be the same low value as given in the table but the flooring contractor should be consulted as to suitable timbers. Also consult the TRADA Advisory Service leaflet No. 1 which lists 9 suitable species of timber.

Concerning screeded subfloors the addition of waterproofers to the mix is NO guarantee against rising moisture. There should always be a damp-proof membrane between screed and subfloor even where wood block floors are laid in mastic. Reference should be made to BRS Digest 104, 'Floor Screeds'.

On suspended ground floors underfloor ventilation is essential. In these cases and at upper floors softwood boarding usually combines the function of structural flooring and finish. Hardwood in the form of strips, parquet or similar is usually laid on an under-floor of softwood boarding, plywood sheets or particle board for reasons of economy.

All kiln dried timber should be close-piled, and should be fully protected from the effects of weather in transit.

There are appropriate Codes of Practice for floors, doors, windows and the structural use of timber in building which should be consulted.

The ideal material to apply to timber for the prevention of moisture pick-up has not yet been discovered, but aluminium primer of a leafing quality has been proved very efficient in this respect. Back-priming of woodwork adjacent to or contiguous with brick walls, etc. with either aluminium primer or a wood preservative (paying particular attention to end-grain surfaces) is a wise precaution.

Preservation

Where timber is to be used in situations conducive to decay non-durable species should be treated with a suitable preservative. See Section Four.

Section six: specification of veneers and veneered panels

Decorative veneers

Selection

Except for very simple work, the architect or designer and (if thought desirable) the client should personally see and choose veneer to be used. Only these people know full circumstances and requirements of work and in any case personal preferences (which no supplier can guess at) play a vital part in this selection.

Selection should be made at the earliest possible stage to ensure wood is available in sizes called for, or to enable design to be modified, if necessary, to accommodate veneer desired.

Personal selection is essential because:

i Details of sizes, placing of joints, method of matching, etc., can be discussed and agreed with supplier.

ii Small samples can be highly misleading and should be avoided as a basis for major decisions.

Certain firms specialise in supplying veneers for architectural purposes and a list of these can be obtained from TRADA.

Specification of veneer

This must identify beyond doubt the stock of veneer chosen for the work because:

a Veneers from different logs of the same species can vary enormously in appearance. To specify 'Walnut Veneer' is quite useless – unless the final appearance is of no consequence.

b With veneer thus specified all panel-makers tendering for the work are basing their prices on the material chosen and the architect/designer knows that the appearance of the work will be the same, so far as the veneer is concerned, regardless of who received the order. Individual interpretations of 'Walnut', for example, are thus eliminated.

A suggested form of specification is as follows:

The panels (doors etc. etc.) are to be faced on one or both sides		
as required with	(name)	veneer
stock number	(Supplier's stock number of approved stock)	
to be supplied by	(name of supplier)	

Types of decorative figure available

There are many ways of describing decorative figure in a veneer. Some of these are plain, ribbon or broken stripe, rope, roe, mottle, fiddle-mottle, raindrop, curly, blister, birds eye and burr.

Veneered panels

Specification

Specifications calling for panels veneered with decorative woods should be carefully drawn up and a clear distinction made between:

a veneered panels to be taken from stock

b plywoods, particle boards or other cores to be veneered to specification.

Veneer stock panels. In this case it will be sufficient for the enquirer to give the thickness, size, quality, quantity and nature of the face veneer required, since the decorative face is in this case an integral part of the make-up of the panel.

If it is desired to make a selection of some particular type of figure then this should be included in the specification. Such a specification will indicate that the tender must allow for the additional price to cover the expense of opening bundles and turning over stock to obtain the required material.

Plywood, particle boards or other cores to be specially veneered. In this case it is necessary to specify the type, thicknesses and dimensions of the core material. Design and arrangement of the veneers should be detailed and provision made for a compensating veneer to be added to the back.

It should be specified whether the face veneers are to be sanded or unsanded.

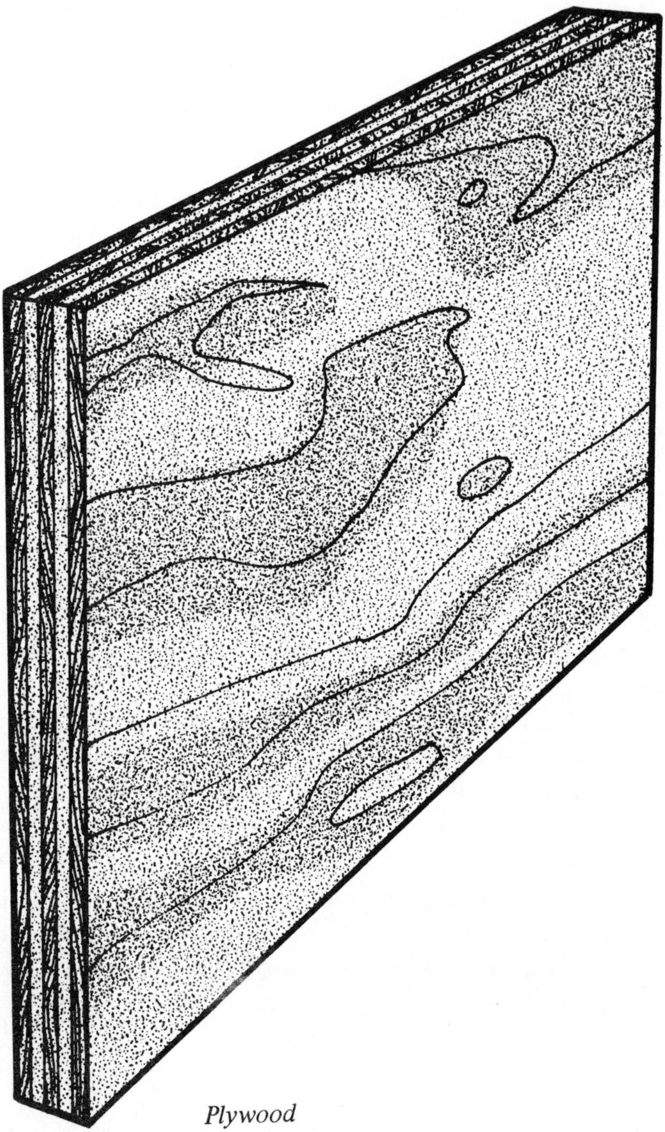

Plywood

Section seven: specification of plywood

General specification

It should be emphasised at this point that when specifying a certain grade of plywood this refers to the quality of the face veneers and bears no relationship to the type. Where specific adhesive standards are required reference should be made to British Standard 1203:1954.

For complete specification the following information should be given to the supplier to enable him to suggest the most suitable plywood to meet specific requirements:

a Thickness
b Species
c Quality
d Size
e Bond quality
f Construction
g Purpose for which required
h Method of finishing
i Any special requirements

Selection

The selection of plywood will depend on its end use. Where appearance is of prime importance then the grade of plywood selected will be the important factor. If both sides are exposed to view then a high quality on both sides will be required. For clear finishing or a high grade paint finish, first quality should be selected.

Where strength is the over-riding factor then, thickness, arrangement and number of plies are important.

The stability is also increased in direct ratio to the number of plies. The situation in which it is to be used will determine whether the bond has to be of internal, external or an intermediate type.

For certain purposes special plywood construction can be obtained. In cases where decorative hygienic surfaces, easy to clean, are required then metal or plastic faced plywood could be specified. For heat insulation plywood with cores or cork, asbestos fibre, foam rubber, etc. can be made to order.

Costs and grades

There are many factors which affect the cost of plywood and it is felt that this publication could not deal adequately with the subject. It is suggested that current prices be obtained from local plywood merchants.

Grading rules for plywood are dependent on country of manufacture and species. In many cases no international standard exists and each manufacturer may have his own specification for quality, gluing, etc. which is notified to his customer in advance. Generally speaking the grades of plywood from any one country conform fairly closely. Appendix 'E' has been compiled to give some guidance to the buyer. The adhesives used in the manufacture of imported plywoods vary, some being suitable for internal use only, while others can be used externally. Details of grades, species and adhesive types available will be found in Appendix 'E' p.119. 'Plywood Species, Sizes and Grades'.

Sizes of plywood are usually given in the standard size to which the boards are manufactured. Larger sizes can be obtained to order by having the material scarf jointed.

When ordering plywood it should be noted that the first dimension quoted indicates the length of the board, i.e. the measurement parallel to the grain of the face veneer. Thus a board 2440 mm x 1220 mm has the grain of the face veneer parallel to the longer dimensions. Conversely a board 1220 mm x 2440 mm is known as a cross grain board, with the grain of the face veneers parallel to the smaller dimension, which is often specified for veneering or bending purposes. Contrary to European practice Canada and USA may quote the first dimension to indicate the width.

If further information is required on grading of plywood, reference should be made to the Timber Research and Development Association publication 'Plywood'.

Section eight: specification of block-board and laminboard

General specification

A complete specification for purchasing blockboard and laminboard should include the following:

a Purpose for which required
b Species of face veneer
c Thickness
d Dimensions
e Type of bond
f Construction of board
g Special requirements

Selection

Blockboard and laminboard are normally manufactured for interior use.

Generally blockboard is used for panelling, partitions, doors, and other interior fitments and it can be finished by painting, veneering or left in the natural state. When a high gloss finish is applied to a board which is going to be viewed under oblique lighting surface irregularities may become apparent. It is of importance that in this case the manufacturer is informed of its use in order that precautions may be taken to prevent this occurring.

Blockboard is particularly suited for covering large areas where depth of section and stability are important. In this respect it is superior to plywood and can be obtained in longer standard lengths. A high class laminboard has all the advantages and superior properties of blockboards and should be used in preference where a superior finish is required both for painting and veneering.

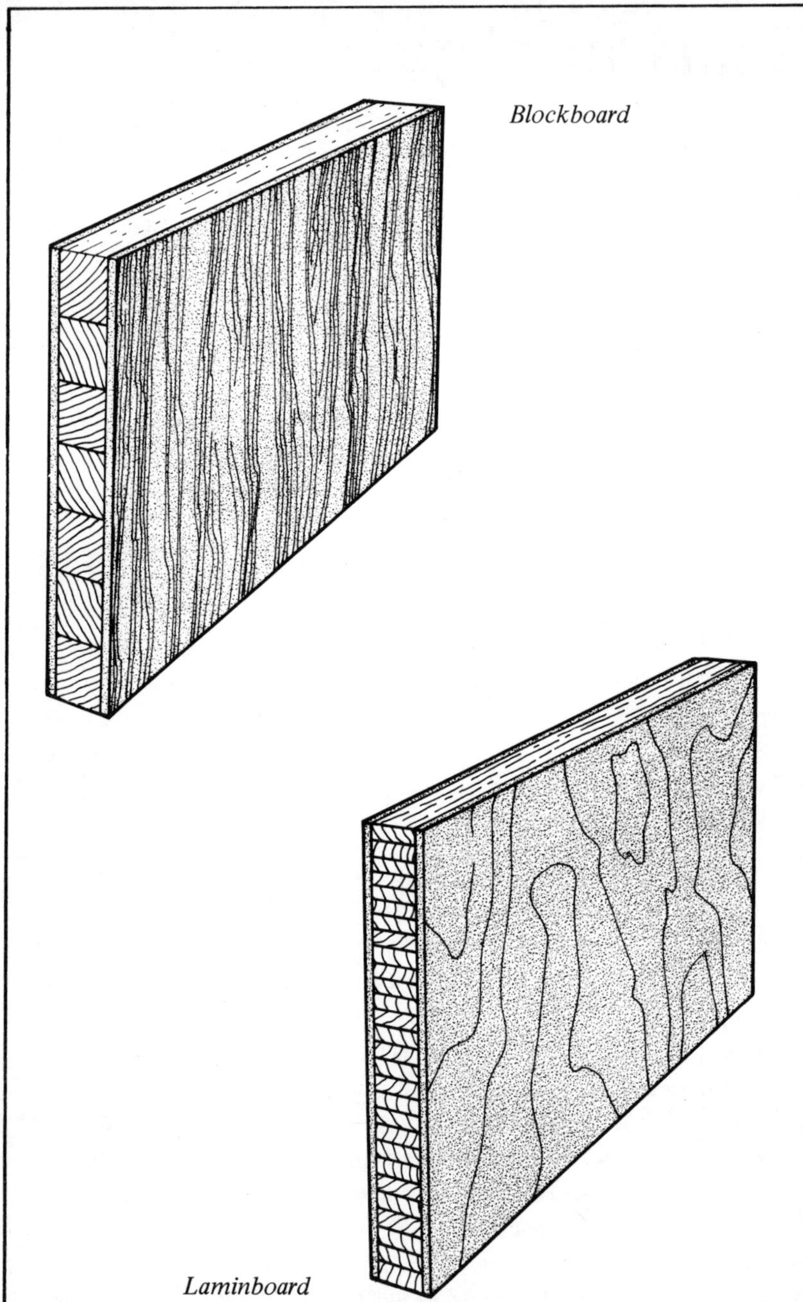
Blockboard
Laminboard

Costs and grades

The cost will vary according to sizes and thicknesses, core species and face veneers. It is therefore, felt that the buyer should obtain current prices from local merchants. In general blockboard is less expensive than plywood of comparable thickness whereas laminboard is more expensive.

Blockboard and laminboard are manufactured with two grades of surface veneer. These are 1st quality face veneer and 2nd quality back (I/II) or 2nd quality veneer on face and back (II/II) or BB.

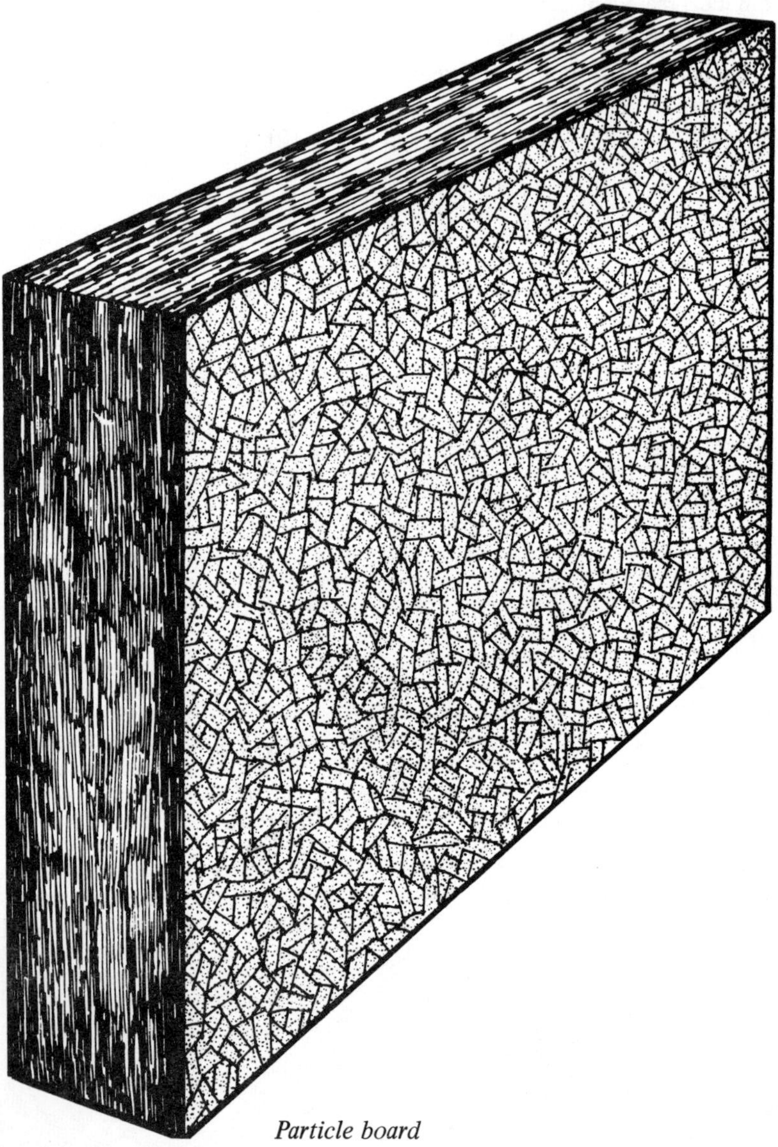

Particle board

Section nine: specification of particle boards

General specification

Particle board is made from specially prepared particles of lignocellulosic materials such as wood, flax, or hemp, bonded with synthetic resins or other organic binders.

All particle boards generally conform to the requirements of BS 2604 'Resin-bonded Wood Chipboard' in which is given requirements for dimensions and tolerances, freedom from foreign particles, moisture content, thermal conductivity, surface spread of flame and minimum strength properties. Test and test methods for wood chipboard and other particle boards are specified in BS 1811 which also incorporates an addendum relating to flooring grade boards.

Particle boards are manufactured by highly developed mechanical techniques which ensure strict regularity of size, thickness, moisture content and other physical characteristics. Thus large flat panels are produced which have no direction of grain and are free from defects.

Application

Particle boards can be used for a wide variety of applications such as roof-decking, flooring, wall-lining, partitioning, built-in fitments and other joinery such as door-cores. Within these applications it is possible to obtain boards of specific nature and characteristics, e.f. flooring grade boards; ready-to-paint grades; boards with fine surface texture; fire resistant and fire retardant boards; boards faced with wood, plastic and hardboard veneers; low-density panels (usually flaxboard) for roof-decking etc.

For new building, particularly industrialised techniques, architects are generally recommended to consider the use of particle board early in their designs in order to obtain full benefit from the material.

The thermal conductivity ('K' value) of wood particle board at 0.144 $W/m^0 C$ is similar to that of most softwoods. The 'K' value of flax based particle board is even better at 0.072 $W/m^0 C$. In situations where both mechanical strength and thermal insulation are required particle boards have much to recommend them.

Storage on site

What has been said about storage on site for joinery etc., (page 36) is of equally vital importance for particle board. In any case it should not be left on site too long prior to use and should be stored flat.

Costs and types

Particle board being a manufactured material is not subject to grading for natural defects carried out for plywood or solid timber. It is obtainable in varying densities which affect strength and thermal properties and determine the purpose for which it is used. Generally speaking on a cost per square metre basis it is cheaper than other comparable materials.

Particle Boards are manufactured in many countries including the U.K.

The following countries are the principal sources of supply of imported boards:-

WOOD CHIPBOARD

Bulgaria
Czechoslovakia
Finland
West Germany
Irish Republic
Israel
Norway
Poland
Portugal
Rumania
Soviet Union
Sweden

FLAXBOARD

Belgium
Czechoslovakia
France
Poland

Appendix 'I' gives an indication of the sizes available.

The Boards are usually marketed under proprietary brand names and since cost and properties will vary from one type of board to another it is suggested that the manufacturer or supplier be approached for the required information on a particular brand. Lists of brand names of boards with comprehensive details relating to qualities, sizes, etc., and information regarding suppliers, can be obtained from the following commercial publications:-

"Building Board Directory"
The Middlesex Publishing Co. Ltd.,
194-200 Bishopsgate,
London, E.C.2.

"Board - Purchasing Guide & Directory"
Pressmedia Limited,
Ivy Hatch,
Sevenoaks, Kent.

Appendix A

Classification and uses of timber in building according to colour

Colour

The list of timbers has been grouped according to colour. The divisions are somewhat arbitrary and most timbers show variation in colour as well as change of colour in use. Most timbers darken with age and particularly under the action of light; some, however, are bleached.

Uses

The list is not exhaustive, and most timbers can be used far more widely than has been listed.

Weight

The density of timber varies with different species and with the moisture content. The values quoted are at 12 per cent but the increase caused by moisture can be calculated by adding 0.5 per cent of the given weight for every 1 per cent increase of moisture content.

Moisture movement

This term refers to the dimensional changes that take place when seasoned timber is subjected to changes in atmospheric conditions (see pages 18, 34 and 66).

Supply

The supply position is liable to change from time to time and enquiries should be made before specifying large quantities.

Price range – key

The following terms are used in the table to indicate prices.
High = over £100 per cu.m.
Medium = £50 to £100 per cu.m.
Low = up to £50 per cu.m.
This range of prices is based on parcels not less than 1.5 cu.m. 'ex yard' of a normal fair average specification at 1970 price levels.

Classification and uses of timber in building according to colour

Colour	Common and botanical name and country of origin	Uses	Texture	Density Kg/m^3	Working qualities	Moisture movement	Supply	Price range	Remarks
White group									
Whitish	**Ash** *Fraxinus excelsior* Europe	Internal Joinery	Medium	689	Good	Medium	Fair	Medium	Obtainable as veneer and plywood. Occasionally has darker markings. American ash is dearer
Whitish	**Birch** *Betula* spp.	Flooring, Internal Joinery	Fine	657	Good	Large	Fair	High	Sizes available should be checked
Whitish	**Hornbeam** *Carpinus betulus* Europe		Fine	753	Difficult	Large	Short	Medium	
Whitish	**Lime** *Tilia vulgaris* Europe	Mouldings	Fine	545	Good	Medium	Short	Low	
Whitish	**Maple, Rock** *Acer saccharum* Canada & USA	Flooring, Panelling	Fine	721	Medium	Medium	Fair	Medium	Available as veneers and also plywood. Several species; may also show colour markings

Classification and uses of timber in building according to colour (continued)

Whitish	**Sycamore** *Acer pseudo-platanus* Europe	Internal Joinery, Panelling	Fine	609	Good	Medium	Fair	Medium	Available as decorative veneers
Whitish	**Whitewood European** *Picea abies* Europe	Internal Joinery, Flooring, Skirting	Fine	465	Good	Small	Good	Low	Can be used panelling in as T. & G. boards
Yellowish	**Afara/Limba** *Terminalia superba* W. Africa	Internal Joinery	Medium	545	Good	Small	Good	Medium	Used for core veneers and occasionally face in construction of plywood
Yellowish	**Jelutong** *Dyera costulata* Malaya	Internal Joinery, Door stock	Fine	465	Good	Small	Fair	Low/ Medium	Latex tubes are present as oval slits – these eliminate its use as a decorative timber
Yellowish	**Obeche** *Triplochiton scleroxylon* W. Africa	Internal Joinery	Medium	384	Good	Small	Good	Low	The timber is used for plywood but is soft. Suitable for veneering on
Yellowish	**Podo** *Podocarpus* spp. E. Africa	Internal Joinery, Flooring	Fine	513	Good	Medium	Short	Low	Has been used for plywood

Classification and uses of timber in building according to colour (continued)

Colour	Common and botanical name and country of origin	Uses	Textures	Density Kg/m^3	Working qualities	Moisture movement	Supply	Price range	Remarks
Yellowish	**Ramin/ Melawis** *Gonystylus* spp. Sarawak and Malaya	Internal Joinery, Flooring, Mouldings	Medium	657	Medium	Large	Good	Low/ Medium	
Brown group									
Light Brown	**Abura** *Mitragyna ciliata* W. Africa	Internal Joinery, Mouldings	Medium/ Fine	561	Medium	Small	Good	Low	Variable in colour with a tendency to be pinkish
Light Brown	**Afrormosia** *Afrormosia elata* W. Africa	Internal Joinery, External Joinery, Flooring, Curtain Walling	Medium/ Fine	689	Medium	Small	Good	Medium/ High	Tends to darken on exposure — will stain in contact with iron under damp conditions. Can also be used as cladding
Light Brown	**Agba** *Gossweiller-odendron balsamiferum* W. Africa	Internal Joinery, External Joinery, Flooring, Curtain Walling	Medium	513	Good	Small	Good	Medium	Gum exudation may be troublesome. Can also be used as cladding

Classification and uses of timber in building according to colour (continued)

Light brown	**Beech** *Fagus* spp. Europe and Japan	Flooring, Internal Joinery	Fine	721	Good	Large	Good	Low	Widely used for plywood. Steamed beech is pink in colour
Light brown	**Chestnut, Sweet** *Castanea Sativa* Europe	Joinery	Medium	545	Good	Small	Fair	Low	Used sometimes as alternative to oak – will stain in contact with iron under damp conditions
Light brown	**Elm** *Ulmus* spp. Europe, USA and Japan	Internal Joinery	Medium/ Coarse	625 average	Medium	Medium	Fair	Low/ Medium	Available as veneers, with careful selection could receive greater attention as a decorative wood
Light brown	**Oak** *Quercus* spp. Europe, America and Japan	Internal Joinery, External Joinery, Flooring	Medium/ Coarse	625–721	Medium/ Difficult	Medium	Good	Medium/ High	Jap. oak paler in colour, milder, available as plywood. Decorative veneers
Light brown	**Olive, East African** *Olea hochstetteri* E. Africa	Flooring	Fine	881	Difficult	Small	Fair	Medium/ High	Only as flooring blocks. Shows wavy dark markings. Loliondo is similar but slightly pinker

Classification and uses of timber in building according to colour (continued)

Colour	Common and botanical name and country of origin	Uses	Textures	Density Kg/m^3	Working qualities	Moisture movement	Supply	Price range	Remarks
Light brown	**Seraya, White** *Parashorea* spp. Borneo	Flooring, Internal Joinery	Medium	529	Medium	Small	Fair	Low	Available as a plywood
Light brown	**Cedar, English** *Cedrus* spp. England	Internal Joinery	Fine	561	Good	Medium	Short	Low	Decorative veneers — is an attractive scented timber. Sizes available should be checked
Light brown	**Hemlock, Western** *Tsuga heterophylla* Canada & USA	Internal Joinery, Flooring	Fine	481	Good	Small	Good	Low	Possesses an attractive lustre. Could be used for panelling and ceiling in T. & G. board form. Can also be used as cladding if preserved
Light brown	**Pine, Yellow** *Pinus strobus* E. Canada	Internal Joinery, Moulding, Panelling	Fine	384	Good	Small	Fair	High	Sometimes pinkish

Classification and uses of timber in building according to colour (continued)

	Teak *Tectona grandis*, Burma and Siam	Internal Joinery, External Joinery, Curtain Walling, Flooring	Medium	641	Medium	Small	Fairly Good	Medium/ High	Available as veneers. Extremely valuable and stable
Reddish brown	**Afzelia/ Doussie** *Afzelia africana* and spp. W. & E.Africa	Internal Joinery, External Joinery, Curtain Walling, Flooring, Cladding	Medium/ Coarse	817	Medium/ Difficult	Small	Short	Medium	Pre-boring is necessary to prevent splitting when nailing – not easy to glue. Exudes a yellow dye in contact with water
Reddish brown	**Cedar, Central and South American** *Cedrela* spp. Central & S. America	Internal Joinery, Panelling	Medium/ Coarse	481	Good	Small	Short	Medium/ High	
Reddish brown	**Cherry** *Prunus avium* Europe and Japan	Internal Joinery	Fine	609	Medium	Medium	Short	Low	Sizes available should be checked
Reddish brown	**Danta** *Nesogordonia papaverifera* W. Africa	Flooring	Medium	737	Good	Medium	Short	Low/ Medium	Sometimes may be dark brown

Classification and uses of timber in building according to colour (continued)

Colour	Common and botanical name and country of origin	Uses	Textures	Density Kg/m^3	Working qualities	Moisture movement	Supply	Price range	Remarks
Reddish brown	**Gedu Nohor/ Edinam** *Entandrophragma angolense* W. Africa	Internal Joinery, Flooring	Medium	545	Medium	Small	Fairly Good	Low	Available as decorative veneers
Reddish brown	**Guarea** *Guarea* spp. W. Africa	Internal Joinery, External Joinery, Cladding, Flooring	Medium	577	Good	Small	Good	Medium	Plywood and veneers
Reddish brown	**Apitong/ Gurjun/ Keruing/ Yang** *Dipterocarpus* spp. India, Malaya, Siam Philippines	Internal Joinery, External Joinery, Curtain Walling, Cladding, Flooring, Structural	Medium	721 average	Difficult	Large	Good	Low	Gum exudation can cause trouble. Used extensively for structural purposes
Reddish brown	**Kapur** *Dryobalanops* sp.	External Joinery, Internal	Medium	737	Medium	Medium	Short	Low	Also known as Borneo camphorwood –

Classification and uses of timber in building according to colour (continued)

	Malaya Borneo	Joinery, Flooring, Curtain Walling							stains in contact with iron. It is almost impossible to obtain unaffected by pin-hole bores although this does not impair its strength properties
Reddish brown	**Mahogany, African** *Khaya* spp. W. Africa	Internal Joinery, External Joinery, Cladding, Flooring	Medium	705	Medium	Small	Good	Medium	Available as plywood and veneers
	Mahogany, American *Swietenia macrophylla* Central and South America	Internal Joinery, External Joinery, Panelling	Fine/ Medium	545	Good	Small	Fairly Good	High	Decorative veneers. On exposure to sunlight becomes golden as do most mahoganies
Reddish brown	**Niangon/ Nyankom** *Tarrietia utilis* W. Africa	Internal Joinery, External Joinery, Cladding, Flooring	Medium	625	Good	Medium	Fairly Good	Medium	Apt to be gummy

Classification and uses of timber in building according to colour (continued)

Colour	Common and botanical name and country of origin	Uses	Textures	Density Kg/m^3	Working qualities	Moisture movement	Supply	Price range	Remarks
Reddish brown	**Plane** *Acer pseudo-platanus* Europe	Mouldings, Internal Joinery	Fine	625	Good	Medium	Short	Medium	Available as veneers. Known also as lace-wood — in Scotland as sycamore. Sizes available should be checked
Reddish brown	**Sapele** *Entandro-phragma cylindricum* W. Africa	Internal Joinery, External Joinery, Curtain Walling, Cladding, Flooring	Fine/ Medium	625	Medium	Medium	Good	Medium	Available as plywood and veneers. Sills
Reddish brown	**Utile** *Entandro-phragma utile* W. Africa	Internal Joinery, External Joinery, Curtain Walling, Cladding, Flooring	Medium	657	Good	Medium	Good	Medium	As for sapele

Classification and uses of timber in building according to colour (continued)

Pinkish brown	**Douglas fir** *Pseudotsuga taxifolia* Canada and USA	Internal Joinery, External Joinery, Curtain Walling, Cladding, Flooring	Fine/ Medium	529	Good	Small	Good	Low/ Medium	As plywood available in decorative forms. Flooring should be rift sawn
Pinkish brown	**Pine, Pitch** *Pinus palustris, P. caribaea* and spp. USA and Central America	External Joinery, Internal Joinery, Curtain Walling, Flooring, Cladding	Fine/ Medium	689 average	Good	Medium	Fair	Low/ Medium	A constructional timber but is also used decoratively
Pinkish brown	**Pine, Scots/ Redwood** *Pinus sylvestris* Europe	Internal Joinery, External Joinery, Cladding, Flooring	Fine/ Medium	513	Good	Medium	Good	Low	Requires protection when used externally as cladding and joinery. Can also be yellowish in colour
Dark brown	**Black Bean** *Castanospermum australe* N.S. Wales & Queensland	Internal Joinery	Medium	705	Medium	Medium	Short	High	Chiefly available as veneer

Classification and uses of timber in building according to colour (continued)

Colour	Common and botanical name and country of origin	Uses	Textures	Density Kg/m^3	Working qualities	Moisture movement	Supply	Price range	Remarks
Dark brown	**Ekki** *Lophira alata* W. Africa	Structural	Coarse	1025 Average	Difficult	Medium	Short	Medium	Heavy construction wood
Dark brown	**Imbuya** *Phoebe porosa* S. American	Internal Joinery, Panelling	Fine/ Medium	721 Average	Good	Small	Short	High	Also known as Brazilian walnut — available as veneers
Dark brown	**Laurel, Indian** *Terminalia* spp. India & Burma	Internal Joinery, Panelling	Medium	849	Difficult	Medium	Short	High	Available as veneers — streaked with darker markings
Dark brown	**Iroko/Mvule** *Chlorophora excelsa* W. and E. Africa	Internal Joinery, External Joinery, Curtain Walling, Cladding, Flooring	Medium	641	Medium/ Difficult	Small	Good	Medium/ Low	Initially lightish yellow but darkens to deep brown — available as veneers
Dark brown	**Muninga** *Pterocarpus angolensis* E. Africa	Internal Joinery, External Joinery, Flooring	Medium	625	Good	Small	Short	Medium	Available as veneers

Classification and uses of timber in building according to colour (continued)

Red group									
Light red	**Makore** *Tieghemella heckelii* W. Africa	Internal Joinery, External Joinery, Curtain Walling, Cladding, Flooring	Fine	625	Medium/ Difficult	Small	Fairly Good	Medium	Available as plywood and veneers
Light red	**Meranti/ Seraya** *Shorea* spp. Malaya and Borneo	Internal Joinery, External Joinery, Cladding, Flooring	Medium	529 Average	Good	Small	Fairly Good	Low/ Medium	Available as plywood. Varies from light to dark in colour. Darker material heavier and more durable
Dark red	**Jarrah** *Eucalyptus marginata* Australia	Flooring, External Joinery, Curtain Walling	Medium	801	Difficult	Medium	Fairly Good	Medium	Occasionally figured veneers
Dark red	**Padauk** *Pterocarpus* spp. Burma and Andamans	Internal Joinery, External Joinery,	Medium	737 Average	Difficult	Small	Short	Medium/ High	Andaman padauk available as veneers
Yellow group									
	African 'Walnut' *Lovoa*	Internal Joinery, External	Medium	545	Medium	Small	Good	Medium	Has irregular dark lines, tending to

Classification and uses of timber in building according to colour (continued)

Colour	Common and botanical name and country of origin	Uses	Textures	Density Kg/m^3	Working qualities	Moisture movement	Supply	Price range	Remarks
	trichiloides W. Africa	Joinery, Cladding, Flooring							brown. Available as veneers
	Idigbo *Terminalia ivorensia* W. Africa	Internal Joinery, External Joinery, Curtain Walling, Cladding, Flooring	Medium	545	Medium	Small	Good	Low	Stains yellow in contact with water
	Opepe *Nauclea diderrichii* W. Africa	Internal Joinery, External Joinery, Curtain Walling, Flooring	Medium	737	Medium	Small	Good	Low/ Medium	Darkens on exposure to a rich reddish brown
Variegated group	Timbers which cannot be easily classified due to the combination of colours in the same piece								
	Blackwood, Australian *Acacia melanoxylon*	Internal Joinery	Medium	657	Medium	Medium	Short	High	Gold or red to dark brown with dark markings. Available as veneers

Classification and uses of timber in building according to colour (continued)

Ebonies *Diospyros* spp. Africa and Far East	Mouldings	Highly decorative material for special purposes but available as decorative veneers						Black streaked with reddish brown markings
Mansonia *Mansonia altissima* W. Africa	Internal Joinery, External Joinery, Flooring	Fine	593	Good	Medium	Fairly Good	Medium	Purple when fresh fading to light fawn on exposure. Dust very irritant
Rosewood *Dalbergia* spp. India and S. America	Internal Joinery	Medium	849 average	Difficult	Small	Short	High	Purplish brown with irregular black markings Decorative veneers
Walnut, European *Juglans regia* Europe	Internal Joinery	Medium	641	Medium	Medium	Fairly Good	High	Decorative veneers. Colour shows considerable variation. Greyish brown with darker markings
Cedar, Western Red *Thuja plicata* Canada and USA	Cladding, Panelling	Medium/ Coarse	368	Good	Small	Fairly Good	Low/ Medium	From light pink to chocolate brown. A very soft timber suitable for wall ceiling linings. Stains with iron under damp conditions

Classification and uses of timber in building according to colour (continued)

Colour	Common and botanical name and country of origin	Uses	Textures	Density Kg/m^3	Working qualities	Moisture movement	Supply	Price range	Remarks
	Parana Pine *Araucaria angustifolia* Brazil	Internal Joinery, Flooring	Fine	545	Good	Large	Fairly Good	Low	Brown to bright red and dark streaks. Available as plywood

Appendix B

Sizes of softwoods

Ordinary building softwoods comprise only about six to eight species and are all imported as sawn stock in a fairly regular range of sizes. Tables 1 – 6 have been prepared to give quick reference to those sizes and with a clear indication of the most easily available sizes, and also the less common dimensions. In using these tables it should always be borne in mind that allowance must be made for shaping and planing.

The distribution of sizes varies geographically in the UK according to the conditions of local trade. It may not, therefore, be possible to obtain all the sizes listed with equal facility in all areas, and it is recommended that local timber suppliers be consulted. For standard sizes see BSS 4471:1969.

Softwoods are imported from the following areas:

Europe

The largest amount of softwood is imported from such European countries as Sweden, Finland and the USSR. Smaller quantities come from Yugoslavia, Czechoslovakia, Poland, France, etc. Generally the trees in Europe are smaller than those in Western Canada and the sawn timber is, therefore, of smaller dimensions. The species mainly concerned are redwood and whitewood and they are very widely used for building: the lower grades for carcassing and the better for joinery.

Widths are never more than 250 mm and the prices of 175 mm and wider are very much higher than those for 150 mm and narrower. Lengths in scantling sizes commonly run up to 5.100 m but are more difficult to obtain and are more costly in longer lengths.

Canada – west coast

Trees from Western Canada are generally much larger than those in Europe and in this part of Canada the Pacific Coast Forest has the largest trees and is the main supplier of Canadian timber to the U.K. Species most widely imported are Pacific Coast Hemlock, Western Red Cedar and Douglas Fir. Sitka Spruce and more recently Western White Spruce are also imported.

Large dimensions of sawn timber are obtained mainly in Pacific Coast Hemlock and Douglas Fir which normally yield lengths up to 9.600 m, although longer lengths are available. Lengths 7.300 m and longer in smaller dimensions are normally obtained from large sizes by re-sawing in the U.K.

Softwoods surfaced four sides

Most of the construction timber used in North America is surfaced four sides and this is also available to U.K. importers as C.L.S. timber, (i.e. timber fully surfaced to Canadian Lumber Standards).

This timber is supplied in grades of construction including 25% standard and is used for general carcassing purposes in house building and general construction. Sizes 150 mm (6 in) and wider are stress graded.

C.L.S. timber will continue to be supplied in Imperial measure in the range of sizes indicated in the following table. Metric equivalents of the exact sizes are indicated for convenient conversions.

Nominal Inches	2"	3"	4"	6"	8"	10"
Actual Inches	$1\frac{9}{16}$	$2\frac{9}{16}$	$3\frac{9}{16}$	$5\frac{5}{8}$	$7\frac{1}{2}$	$9\frac{1}{2}$
Metric Equivalent (mm)	39.7	65.1	90.5	142.9	190.5	241.3

Canada – eastern

Spruce is the timber imported; it resembles European Whitewood in appearance and properties and is available in rather similar sizes.

Other countries

Considerable quantities of Parana pine are imported from Brazil. This timber has the advantage of being normally available in widths up to 360 mm.

Although not so readily available as in the past, pitch pine is still imported from the Southern States of the USA and also from British Honduras, and smaller material from the Bahamas.

Table 1. Redwood and Whitewood.

Thickness mm	*Width* *75*	*100*	*125*	*150*	*175*	*200*	*225*	*250*	*300*
16	O	N	N	N					
19	N	N	N	N					
22	N	N	N	N					
25	N	N	N	N	N				
32	N	N	N	N	N				
38	O	N	N	N	N	O	N		
44	O	O	O	O	O	O	O	O	O
50	N	N	N	N	N	N	N	O	O
63		N	N	N	N	N	N		
75		N	N	N	N	N	N	O	O
100		O		N		N		O	O
150				O		O			O
200						O			
250								O	
300									O

Lengths Beginning at 1.800 m increments by 300 mm.
Range 1.500 m – 6.300 m
Occasionally greater lengths are available if shippers are prepared to cut to special order.

Table 2. Parana pine.

Dimensions (Imperial Sizes (inches) in Brackets)

	(6) 142.8	**(8)** 190.5	**(9)** 215	**(10)** 245	**(11)** 270	**(12)** 295
($\frac{3}{4}$) 19	O	O	O	O	X	O
(1) 25	N	N	N	N	N	N
($1\frac{1}{4}$) 32	N	N	N	N	N	N
($1\frac{1}{2}$) 38	N	N	N	N	N	N
(2) 50	N	N	N	X	X	N
(3) 75	O	X	O	X	X	O

Lengths (metres)
Mainly 4.800
Occasional 75 mm and 100 mm in strips.

Table 3. Eastern Canada (mainly Spruce).
(Imperial Sizes (inches) in Brackets).

Nominal Sizes mm	**76.2** (3)	**101.6** (4)	**127** (5)	**152.4** (6)	**177.8** (7)	**203.2** (8)	**228.6** (9)	**254** (10)	**279.4** (11)	**304.8** (12)
25 (1)	O	O	O	O	O	X	X	X	X	X
50 (2)	N	N	N	N	N	N	O	O	O	O
63 ($2\frac{1}{2}$)	X	N	N	N	N	N	O	O	O	O
75 (3)	X	N	N	N	N	N	N	O	O	O

Lengths (metres) 2.400 – 5.200 Increments by 300 mm
Mostly 3.300 – 4.500
Some longer lengths available subject to special order with shippers.

Table 4. West Coast of Canada – Sawn timber mainly Pacific Coast Hemlock, Douglas Fir and Western Red Cedar. (Imperial Sizes (inches) in Brackets).

Imperial	**mm**	**(3)** 75	**(4)** 100	**(5)** 125	**(6)** 150	**(7)** 175	**(8)** 200	**(9)** 225	**(10)** 250	**(11)** 275	**(12)** 300
(1¾)	44	X	N	X	N	O	N	O	N	X	X
(2)	50	O	N	O	N	O	N	O	N	X	O
(3)	75	X	N	O	N	O	N	O	N	O	O
(4)	100	X	N	O	N	O	N	O	N	O	O

Lengths are available in increments of 600 mm in the range of 1.8 m – 9.6 m depending on the species and type of timber concerned.

Appendix C

Sizes of hardwoods

Trouble is frequently caused by selecting a hardwood on its appearance and specifying it for purposes for which it is either unsuitable or unobtainable in the sizes and lengths required. The suitability for a specific purpose can be determined by reference to the appropriate publications, but the sizes are dependent upon the tree from which the material is cut. The information in this Appendix cannot be described as complete but it covers the general sizes of the more common hardwoods used in building and joinery.

The following notes and the table cover the normal sizes available but should a definite size be required which is not mentioned the timber importers' merchant should be consulted before specifying the particular timber. In using the table below it should always be borne in mind that allowance must be made for planing and shaping.

Widths Sawn hardwoods are normally imported in random widths. This is the direct opposite to the position with softwoods where it is common to have a whole parcel of boards of one width. Therefore large quantities of sawn hardwoods in one width may be difficult to obtain and are likely to be expensive in 150 mm and wider. Keruing and teak (*Tectone grandis*) are available in fixed widths. Wide widths can be easily obtained in West African and home-grown species by cutting from the log (see below).

Strips This term is given to small dimension, square-edged stock, generally 25 mm by 75 mm up to 125 mm. They are cheaper than the larger sizes.

Shorts Short lengths of certain timbers are available at reduced cost and are given this name. Knowledge of the actual lengths of shorts may well result in material saving. In most shipments shorts will be under 1800 mm in length.

Log sizes usually start from 1.8 m rising in increments of 100 mm.

Shorts regularly imported in certain species begin at 1.700 m falling in decrements of 100 mm. Note that in view of the wide gap between the metre and the millimetre some firms may adopt the centimetre as an intermediate measure for hardwood.

The waste factor due to the nature of the species and the grade of logs should be borne in mind when assessing the dimensions of square-edged timber likely to be cut from them.

Countries of origin

Practices with regard to the conversion of logs vary in different countries. In some cases the general run of log sizes is also fairly consistent in one country, e.g. logs in East Africa tend to be smaller than those in West Africa. It is often, therefore, helpful to know the country of origin of a timber and the characteristics of sizes and methods of conversion in the various countries.

West Africa This is one of the few areas from which logs are usually converted in Britain. Much of this conversion is through-and-through, although quarter-sawing can be undertaken, if logs are large enough.

There is a tendency for the amount of through-and-through cutting to 25 mm material to have lessened in recent years. Conversion in this country also means that a very wide range of sizes is available, if the sawmiller is given enough time and the quantities wanted are reasonably large.

It should be remembered that the majority of logs are imported in lengths of 3.600 m to 5.400 m, anything longer is imported separately at higher cost.

Malaya All timber from Malaya is imported as sawn stock. The principal timbers concerned are keruing and red meranti. The bulk of this is 150 mm – 200 mm wide with lengths 1.800 m – 6.000 m averaging 3.600 m – 4.200 m.

Japan Japanese oak is the principal timber imported, with some quantities of such species as elm, maple and beech. All timber is sawn square-edged. It is a fairly general rule in Japanese oak that the wider widths have short lengths and vice versa.

Europe Beech and oak comprise the main hardwoods imported from European countries (principally France and Yugoslavia). All this timber is sawn, although a fairly large proportion is unedged. Both timbers average about 180 mm to 190 mm in width for square-edged materials and 1800 mm to 3000 mm in length. Shorter boards (1.700 m – 0.900 m), known as 'shorts' are available at lower cost. Unedged beech and oak can be obtained in wider sizes.

United Kingdom All home-grown timber is, of course, available in log form in this country; much of it is, however, cut into 25 mm and thicker boards and the sizes vary considerably with species. The trade also tends to be somewhat specialised; one firm dealing mainly in timber for fencing, gates, etc., another in wood for pitprops, a third in oak for church work, etc. It is sometimes difficult therefore, to find a source of the exact material needed; in such cases it is advisable to get in touch with the Federated Home Timber Associations, 68 Cannon Street, London EC4 who will advise as to the best sources of supply.

North America Hardwoods from Canada and the United States of America are seldom imported in lengths of over 4.800 m; usually 2.400 – 4.800 m, and rising in even 600 mm lengths, e.g. 2.400, 3.000, 3.600 m, etc. See also in the table under American oak, Canadian yellow birch and rock maple.

Special uses

Flooring hardwoods A very large number of hardwoods are available for flooring and information about the species and their fitness for particular kinds of flooring can be obtained from the TRADA manual 'Wood Floors'. In addition the FPRL Bulletin No. 40 gives similar data and this is obtainable from HMSO.

Sill timbers Some suitable hardwoods, such as oak, keruing, utile and sapele are often obtainable in sill sizes, i.e. 50 mm x 125 mm, 75 mm x 200 mm.

Sizes in hardwoods

Name	Origin	Available Sizes in mm	Lengths in m	Remarks
Abura	W. Africa	Logs: 400 – 750 diam SE: 150 & up to average 200 – 225	Logs: 3.600 – 9.100 SE: 1.800 & up (av. 2.400 – 2.700)	Limited quantities of SE available
Afara/Limba	W. Africa	Logs: 600 – 900 diam SE: 150 – 250 wide 25 – 75 thick	Logs: 3.600 – 7.300 SE: 1.800 – 5.500	
African 'Walnut'	W. Africa	Logs round: up to 1250 diam Logs squared: 550 x 550 & up SE: 150 – 350 wide 25 – 100 thick	Logs: 3.600 – 6.700 & longer Logs: 3.600 – 5.500 SE: 1.800 – 3.600 0.900 – 1.700	
Afrormosia/Kokrodua	W. Africa	Logs: up to 1500 diam SE: 150 – 450 wide 25 – 100 thick	Logs: 3.600 – 11.000 SE: 2.400 – 6.100	
Afzelia	W. & E. Africa	Logs: 600 – 1000 diam SE: 150 & up 25 – 100 thick	Logs: 3.600 – 7.300 SE: 1.800 – 5.500	
Agba	W. Africa	Logs: 600 – 1800 diam SE: 150 – 450 wide 25 – 100 thick	Logs: 3.600 – 7.300 SE: 1.800 – 5.500	
Ash	Europe	Logs cut through & through	1.800 – 5.500	SE available according to sizes of logs
Beech (French & Rumanian)	Europe	SE: 100 & up wide 25 – 100 thick Unedged: 150 & up wide	1.800 – 3.000 Shorts 1.000 – 1.700 Ultra shorts also available	Also sawn squares Narrower strips available
Beech (Other European)	Europe	SE: 125 & up wide (av. 190) Unedged: 150 & up wide	As above	As above

Sizes in hardwoods (continued)

Birch Canadian Yellow	Canada	Strips: 25 x 75 – 100 SE: 150 & up (av. 225)	2.400 & up	
Cedar	Central & South America	SE: 100 & up wide 25 – 75 thick 25 – 100 thick	1.200 & up mainly 1.800 & up (av. 3.000 – 3.300)	
Chestnut, Sweet	Europe	SE: 25 – 75 thick 100 – 200 wide	1.800 & up	Logs cut through and through available
Elm	Europe, U.S.A. Japan	SE: 19 – 150 & up wide 25 – 100 thick	1.800 & up	
Gedu Nohor/Edinam	W. Africa	Logs: 860 – 1800 diam SE: 150 – 350 wide 25 – 100 thick	Logs: 3.600 – 9.100 SE: 1.800 – 6.100	
Guarea	W. Africa	Logs: 600 – 1250 diam SE: 150 & up wide 25 – 50 thick Strips: 25 x 75 – 100	Logs: 3.600 – 9.100 SE: 1.800 & up Strips	
Gurjun & Yang	Malaya & Siam	Logs: 300 – 700 diam SE: 150 – 380 wide 35 – 150 thick Strips: 25 – 32 x 75 – 100	Logs: 3.600 – 12.100 SE: 1.800 – 7.300	
Idigbo/Emeri	W. Africa	Logs: 600 – 1150 diam SE: 150 – 450 wide 25 – 100 thick	Logs: 3.600 – 10.000 SE: 1.800 – 5.500	
Iroko	W. Africa	Logs round: 600 – 1800 diam Logs square: 450 x 450 & up SE: 150 – 600 & wider 25 – 100 thick	Logs round: 4.200 & up Logs square: 3.600 & up SE: 1.800 – 6.100	
Jarrah	W. Australia	SE: 25 – 150 thick 75 – 300 wide	1.500 – 8.500	Also available as manufactured flooring

Sizes in hardwoods (continued)

Name	Origin	Available Sizes in mm	Lengths in m	Remarks
Jelutong	Malaya	SE: 25 – 100 thick 150 & up wide	up to 4.200	
Kapur	Borneo	Few logs: 450 – 900 diam SE: 150 – 300 wide 25 – 100 thick	Logs: 3.600 – 9.100 SE: 1.800 – 7.300	
Karri	Australia	Flooring strips: 25 x 75 & 25 x 100	1.800 & up	
Keruing	Malaya	SE: 100 & up (mainly 150 – 300 wide) 25 – 150 thick	1.800 – 6.100 & longer	
Laurel, Indian	India & Burma	SE: 25 – 150 thick 100 – 225 wide	1.800 – 4.200	
Lime	Europe	Logs cut through & through 25 – 125 thick	1.800 & up	
Loliondo	E. Africa	Flooring strips		
Mahogany, African	W. Africa	Logs: 600 – 1500 diam SE: 150 – 750 (mainly 150 – 450 wide) 12 – 100 thick Strips: 75 – 125 (mainly 75 & 100 Shorts: 150 & up wide 75 & up wide	SE: 1.800 – 6.100 Occasionally longer (mainly 1.800 – 5.500) Strips: 1.500 & up or 0.900 – 1.800 0.900 & up	
Mahogany, American	Central & South America	150 & up wide (av. 225)	2.400 & up (av. 3.000 – 3.300)	
Makore	W. Africa	Logs: 600 – 1250 diam SE: 150 & up wide mainly 25 – 100 thick Strips: 25 x 75	Logs: 3.600 – 7.300 SE: 1.800 & up Strips: 1.800 & up	

Sizes in hardwoods (continued)

Mansonia	W. Africa	Logs: 400 & up diam SE: 100 & up wide 25 – 75 thick Strips: 25 x 75 – 125	Logs: 3.600 & up SE: 1.800 & up Shorts: 0.900 & up	
Maple, Japanese	Japan	SE: 150 & up wide 25 – 50 thick Strips: 25 x 75 – 125	SE: 1.800 & up Strips: 0.750 – 1.650	
Maple, Rock	Canada & U.S.A.	SE: 150 & up wide 25 – 100 thick Strips: 25 x 75 – 100		Also available as manufactured flooring
Mengkulang	Malaya	SE: 150 & up wide 25 – 100 thick	1.800 – 6.000	
Meranti/Seraya, red	Malaya & Borneo	SE: 100 – 300 wide 12 – 100 thick	SE: 1.800 – 7.300	
Missanda	E. Africa	Short Strips: 25 x 75		
Muhimbi	E. Africa	Strips only: 19 & 25		
Muhuhu	E. Africa	Flooring blocks		
Muninga	E. Africa	SE: 150 – 300 wide 25 – 75 thick Strips: 25 x 75	1.800 – 4.800	
Mvule	E. Africa	Logs round: 600 & up diam Logs square: 250 x 250 & up	Logs round: 3.600 & up Logs square: 2.400 & up	E. African timber tend to be smaller than W. African
Niangon	W. Africa	SE: 25 – 50 thick 150 & up wide Logs: 500 diam & up	1.800 & up Logs: 4.000 – 7.000 long	

Sizes in hardwoods (continued)

Name	Origin	Available Sizes in mm	Lengths in m	Remarks
Oak, American Red & White	U.S.A.	SE: (FAS grade): 150 – 350 (av. 200 – 250 wide) 25 – 100 thick	2.400 & up	Also available as manufactured flooring
		other grades 100 & up (av. 150 – 200) 25 – 100 thick	1.200 – 4.900	
Oak, Japanese	Japan	SE: 150 & up (av. 200 – 212) wide. (wide widths available at increased cost) 12 – 100 thick. Shorts 150 & wider (av. 200 – 212)	1.800 & up (av. 2.400 – 2.600) 0.750 – 1.650	
		Strips: 19 – 25 thick 50 – 125 wide		
Oak, Yugoslavian	Europe	SE: 125 & up (av. 190 wide) 25 – 100 thick	1.800 – 3.000	
		unedged 150 & up wide 25 – 100 thick	1.800 & up	
Oak, other European	Europe	Mainly unedged: 150 & up wide. 25 – 100 thick	1.800 & up	
Obeche	W. Africa	Logs: 450 – 1750 diam (mostly 600 & up)	Logs: 3.600 – 7.600	
		SE: 150 – 550 wide 25 – 100 thick	SE: 1.800 – 5.500	
		Strips: 25 x 75 – 125		
Opepe/Kussia	W. Africa	Logs: 600 – 1250 diam	Logs: 3.600 – 9.100	Also sawn squares
		SE: 150 – 450 wide 25 – 150 thick	SE: 1.800 – 6.100	
		Strips: 25 x 100 & 75	Strips: 1.800 & up & 0.900 & up	

Sizes in hardwoods (continued)

Padauk	Burma and Andamans	SE: 25 – 75 thick 150 – 350 wide	1.800 – 4.900	Ex logs and occasional larger sizes
Panga Panga	E. Africa	Mainly short strips only Some SE: 150 & up wide 25 – 50 thick	SE: 1.800 & up	
Podo (strictly a softwood)	E. Africa	SE: 50 – 300 wide 25 – 75 thick	0.600 – 4.900	
Ramin	Sarawak	Logs: 375 & up diam SE: 150 – 300 wide 19 – 50 thick (mostly 25) Strips: 25 x 50 – 125	Logs: 2.400 – 6.100 (mainly 4.200) SE: 1.800 & up Strips: 1.800 & up & 0.600 – 1.500	
Rhodesian 'Teak'	E. Africa	Flooring blocks only		
Rosewood	India and South America	SE: 19 – 75 thick 75 – 225 wide	1.800 – 2.400	Ex logs and occasional larger sizes
Sapele	W. Africa	Logs: 600 – 1250 diam SE: 150 – 600 wide 25 – 100 thick	Logs: 3.600 – 9.100 SE: 1.800 – 6.100	
Seraya Red & White	Borneo	Logs: 450 – 1000 SE: 150 – 450 wide 12 – 100 thick Strips: 25 x 75 – 100	Logs: 3.600 – 9.100 SE: 1.800 – 7.300	
Sycamore	Europe	Logs cut through & through 19 – 75	1.800 & up	SE available according to sizes of logs
Tasmanian 'Oak'	Australia	SE: 150 & up wide 25 thick Strips: 25 x 75 – 100	1.800 & up	Also available as manufactured flooring

Sizes in hardwoods (continued)

Name	Origin	Available Sizes in mm	Lengths in m	Remarks
Teak	Burma & Siam	SE: Extensive range: 150 – 350 (av. 200 – 225) wide 25 – 100 thick & some over 100 in limited supply. Logs round: 600 & up Logs square: 350 x 350 & up Scantlings: 50 – 75 x 75 – 125 Deckings: 50 – 125 x 100 – 137	SE: 1.800 & up (av. 2.400 – 2.700) Logs round: 3.600 – 6.700 Logs square: 3.600 – 6.100 (av. 4.900) Scantlings: 1.800 & up Deckings: 3.000 – 7.300	
Utile	W. Africa	Logs: 600 – 1800 diam SE: 150 – 600 wide 12 – 100 thick	Logs: 3.600 – 9.100 SE: 1.800 – 6.100	
Walnut, English & French	Europe	Unedged only, widths vary but are small: 25 – 100	Variable, mostly short	

Note: The upper limits of all dimensions must be regarded as very approximate. SE = square edged.

Appendix D

List of decorative veneers

The following list shows some of the timbers which are normally available in veneer form. Full details of these and many other species can be obtained from veneer suppliers for special architectural requirements. A list of suppliers can be obtained from TRADA.

Afrormosia
Agba
Abele
Ash, White
Ash, Olive
Ash, Silver
Aspen
Avodire

Bean, Australian
Beech
Betula, Silver
Birch, Canadian
Birch, Flame
Birch, Masur
Blackbean, Australian
Blackwood, Australian
Bosse
Bubinga
Box

Camphor Laurel
Cedar, Lebanon
Cedar, Indian
Cedar, Marble
Cherry, African
Cherry, American
Cherry, European
Chestnut
Chestnut, Spanish
Courbaril

Daniellia
Douglas Fir

Ebiara
Ebony, Macassar
Elm, European
Eucalyptus

Guarea

Imbuya
Idigbo
Iroko
Indian Silver Greywood

Kevazingo
Kindal

Lacewood
Larch
Laurel, Indian
Limba

Mahogany, African
Mahogany, Burma
Mahogany, Cuban
Mahogany, Honduras
Mahogany, Nicaraguan
Mahogany, Pommele
Makore
Mansonia
Maple, Birds Eye
Maple, Queensland
Maple, Quilted
Maple, Rock
Marillo
Moabi

Myrtle

Oak, American
Oak, English
Oak, French
Oak, Australian Silky
Orangewood

Padauk
Paldao
Peartree, African
Peartree, European
Peroba
Pine
Planetree

Rosewood, Brazilian
Rosewood, Indian

Sapele
Satiney, Red
Satinwood

Sycamore, White
Sycamore, Weathered

Tchitola
Teak
Tulip

Walnut, African
Walnut, American
Walnut, Australian
Walnut, English
Walnut, French
Walnut, Italian
Walnut, Pacific
Walnut, Persian
Wenge

Yellow-wood
Yew

Zebrano

Appendix E

Plywoods: quality and sizes

Grades

Finnish Birch

FACE QUALITIES

B (I) Quality
Allowing a few knots up to 8 mm in diameter, some brown streaks, swirls and slight discolouration. Joints to be glued and adjacent strips matched to colour.

S (II) Quality
Good paintable quality. Knots some open up to 10 mm in diameter and discolouration permitted. Joints are glued.

BB (III) Quality
Knots are plugged except such small ones as are permitted in 'B Quality'; joints and discolouration allowed.

WG (IV) Quality
Only guaranteed to be well glued; admits knots, plugs and joints in any number and some manufacturing defects.

Qualities available

Panels are obtainable in the following combinations of qualities:

	front side	reverse side
B (I) =	B (I)	B (I)
B(BB) I/III =	B (I)	BB (III)
B/WG (I/IV) =	B (I)	WG (IV)
S (II) =	S (II)	S (II)
S/BB (II/III) =	S (II)	BB (III)
S/WG (II/IV) =	S (II)	WG (IV)
BB (III) =	BB (III)	BB (III)
BB/WG (III/IV) =	BB (III)	WG (IV)
WG (IV) =	WG (IV)	WG (IV)

All boards regardless of quality meet the same strength requirements; BB (III) is the most popular for structural work.

Treated boards

Fire Treated FINPLY exterior can be supplied pre-treated by paint, varnish or impregnation to meet Class 1 or alternatively Class 2 surface spread of flame (see Section 2 p.31 and BS. 476 Part 1).

Preservative treatment For permanently damp conditions (moisture content above 20%) boards are obtainable on special order pre-treated with preservatives to give through and through protection from fungi and insect attack. A film faced plywood is available for concrete shuttering (see Appendix 'E').

Plywood floors

FIN-PLY is also available for all kinds of flooring, see TRADA Publication 'Wood Floors'.

A list of importers and further details can be obtained from the Finnish Plywood Development Association.

Russian Birch

Sanded two sides (except C faces)

BB

Well made plugs permitted

CP

Well made plugs and large sound knots permitted

C

All defects allowed; only guaranteed to be well glued

Russian Alder

Sanded two sides (except C faces)

B

Practically free from all defects

BB

Well made, plugs permitted

C

All defects allowed; only guaranteed to be well glued

Note

The following notes refer to three above types of plywood:

1 Certain grades may consist of a combination of the above categories, i.e. B/BB, face side B practically free from all defects, reverse side BB permitting some open defects, etc.

2 Jointed grades are often indicated by the inclusion of the letter J, i.e. BJ/BB.

Swedish Pine

A

A few pearl knots permitted. Some perfect joints permitted.
Surface sanded

B

Some knots or plugs and a few minor defects permitted several perfect joints permitted. Surface sanded.

BB

Several knots, plugs and other defects. Several joints permitted. Surface sanded or scraped.

X

Knots, knot holes, cracks and other defects permitted. Surface sanded or scraped.

Note

1 Combination of grades are imported, i.e. A/B, B/BB, etc.
2. Where 'one good' side is present then the quality of the good side will be denoted by A, B or BB and the reverse by X.
3 Where 'two good' sides are present the qualities will be marked B, B/BB or BB and the grades indicated will apply to the face veneer on both sides.

European Beech

I (A)

Special selection of faces virtually without blemish

II (B)

Some pin knots and discolouration allowed. Jointed unless specified

III (BB)

Small sound knots, plugs and joints allowed. Some streaks and sound discolouration permitted

IV (BBB PQ C)

Permits broken knots, gaps, splits. Rough finish. Well glued

Note

Combination of grades as above, i.e. I/II (A/B), etc.

European Oak

Prime (First)

Special selection of faces, virtually without blemish and of even colour

Second
Some pin knots and discolouration allowed.

Exterior Douglas Fir
Good Two Sides (G25) A/A Ext
Both sides unjointed or if jointed reasonably matched for grain and colour.
Smooth surface
Free from knots, splits, pitch pockets, and other open defects. Streaks, discolouration, sapwood shims and neatly made wood inlays shall be admitted
Good One Side Solid Reverse (G/S) A/B Ext
Face as Good Two Sides above. Reverse shall present a solid surface free from open defects, but in addition to the characteristics admitted on the good face shall admit neatly made repairs including synthetic plugs, knots up to 25 mm if sound and tight, but no torn grain and other minor defects.
Good One Side (G1/S) A/C Ext
Face side as for Good Two Sides. Reverse may contain knot holes up to 25 mm in least dimension, open pitch pockets up to 25 mm, splits up to 5 mm, worm or borer holes up to 16 mm wide or 38 mm long, tight knots up to 38 mm in least dimension, closed patches and sanding defects, may be jointed
Solid Two Sides (Solid 2S) B/B Ext
Both sides similar to the back of Good One Side Reverse
Solid One Side (Solid 1S) B/C Ext
Face side similar to the back of Good One Side Solid Reverse Grade. Reverse side similar to the back of Good One Side grade.
Select Sheathing C plugged C Ext
One face will present a surface suitable for overlay with lino etc. Permits closed patches, sound tight knots, reasonably torn or rough grain, minimum open defects and splits not exceeding 5 mm wide provided that it does not impair the serviceability of the panel. Reverse side is similar to the back of Good One Side.
Regular Sheathing C/C Ext
Both sides similar to the back of Good One Side.
Note
The lettered grades not in brackets above are for Douglas fir plywood manufactured in the USA.

British-made Plywood: British Standard 1455: 1956
The grade of the veneer shall be assessed after the board has been made, and not when in the form of veneer.

Grades of veneer are defined as follows:

Grade 1 veneer

shall be one or more pieces of firm, smoothly cut veneer. When of more than one piece, it shall be well jointed and reasonably matched for grain and colour at the joints. Veneers shall be approximately equal in width, but slight discolouration and an occasional closed split shall be permitted. It shall be free from knots,* worm and beetle holes, splits, dote, glue stains or other defects which would be objectionable when the plywood has to be used in its natural state. No end joints are permitted.

Grade 2 veneer

shall present a solid surface free from open defects. Veneers when jointed need not necessarily be matched for colour or be of equal width. A few sound knots are permitted, with occasional minor discolouration and slight glue stains, isolated pin worm holes not along the plane of the veneer and occasional closed splits of maximum 1/10th of the length of the panel. This grade shall admit neatly made repairs of wood inlays which present solid, level, hard surfaces, and are bonded with an adhesive appropriate to the type of the plywood. No end joints are permitted.

Grade 3 veneer

may include wood defects excluded from Grades 1 and 2 in number and size which will not impair the serviceability of the plywood. It may also include manufacturing defects, such as rough cutting, overlaps, gaps or splits, provided these do not affect the use of the plywood. No end joints are permitted.

Grade S veneer

Specially selected grade to be the subject of agreement between manufacturer and purchaser.

Note

The following are the uses for which plywood made with the grades of veneer defined above are considered suitable:

Grade 1 For use in its natural state.

Grade 2 For use where subsequent painting and/or veneering is intended.

Grade 3 For use where it is not normally visible.

Marine Manufactured to BS 1088, 'British made Plywood for Marine Craft'.

Combination of Grades

The grades of the face and back veneers are assessed separately. The combination of these grades (e.g. 1/2, 2/2, 2/3) defines the grades of the plywood.

**This shall not exclude the occasional 'dormant bud', which resembles a sound knot, not more than 3.2 mm diameter, of the same colour as the surrounding wood. Also permitted are occasional areas of disturbed grain, up to 12.5 mm diameter where dormant buds have been overgrown.*

Japanese plywood

The general description of Japanese plywood manufactured in accordance with the 'Export Standard for Japanese Plywood: 1958' is as follows:

Plywood

The plywood shall be classified into four types, I, II, III, IV. Each type is graded as 1st, 2nd, 3rd or 4th quality. (The grading rules for quality are too complicated to be included here but generally the quality of Japanese plywood has been found satisfactory and superior to that permitted by the grades laid down.)

Types

The four adhesive types, with approximate equivalents added for rough guidance, are:

I water-proof–approximately equivalent to BR
Type I plywood is bonded with melamine, phenol formaldehyde resin or adhesives of equal or superior quality, so that the plywood may withstand some exposure to weather and moisture and be unaffected by micro-organisms

II water-resistant–approximately equivalent to MR
Type II plywood is bonded with urea formaldehyde resin or adhesives of equal or superior quality, so that the plywood may withstand some exposure to weather and moisture and be unaffected by micro-organisms. It is required to pass a cold soaking test and a moisture content test

III moisture-resistant–approximately equivalent to INT
Type III plywood is chiefly for interior use, and is bonded either with an extended urea or casein adhesive. Bonding must be strong enough to resist the effect of slight changes in atmospheric conditions. It is required to pass a cold soaking test, a dry bonding strength test and a moisture content test.

IV non-water proof
Type IV plywood is used where moisture resistance is unimportant. It is bonded with soya bean adhesive or an adhesive of equal or superior quality, and is required to pass a dry bonding test and a moisture content test.

The test requirements are not identical to BS 1203: 54 so that the equivalents are suggested for rough guidance only. Details of the tests are given in an appendix to the 'Export Standard for Japanese Plywood: 1958'.

Indication of types and grades

The following designation letters are stamped on the boards to indicate the type and grade of plywood.

Type	
Type I plywood	P
Type II plywood	D
Type III plywood	T
Type IV plywood	L
Grade	
1st quality	AA
2nd quality	AB
3rd quality	BB
4th quality	CC

Israeli plywood

A/B Quality

A Face unjointed and practically free of defects.

B Face jointed and some small sound defects allowed. Both faces sanded.

B Quality

Face side jointed or unjointed at shipper's option, but jointed veneers are well matched for colour. Some small sound defects are permitted.

Reverse side jointed or unjointed at shipper's option, but veneers are unmatched for colour. Some small defects are permitted. In all cases both faces are sanded.

B/BB Quality

B face unjointed and allowing only minor defects.

BB face permits large sound defects. Boards are only sanded on the face side.

Nigerian plywood

Gold

Face side jointed at shipper's option and reasonably matched for colour, but admitting slight discolouration. Reverse side unmatched for colour and jointed at shipper's option, discolouration, sound knots and occasional short closed splits admitted. Free from wormholes both sides.

Silver

Both sides unmatched for colour and jointed at shipper's option, admitting a reasonable number of defects including isolated wormholes.

Blue

Both sides unmatched for colour and jointed at shipper's option, admitting unlimited defects and isolated wormholes.

Marine

Manufactured to comply with BS 1088, 'British made Plywood for Marine Craft'.

French Gaboon Plywood

A/B Quality

Face unjointed and free from all defects except occasional very small sound knots.

Reverse usually unjointed but some shippers reserve the right to supply jointed faces. Permitting slight discolouration, some small sound knots, small splits and slight roughness.

B Quality

Face and reverse as reverse of A/B Quality.

B/BB Quality

Face as reverse of A/B.

Reverse permitting joints, splits, rough patches, knots, plugs and occasional pin wormholes.

Gaboon plywood from Gabon, West Africa

Premier Grade

Face side jointed or unjointed at shipper's option. When jointed, veneers properly matched for colour. Occasional small veneer defects admitted.

Reverse jointed or unjointed at shipper's option. When jointed, veneers reasonably matched for colour. Occasional small defects admitted. Small open defects, if any, to be well filled. No worm holes.

Standard Grade

Face side jointed or unjointed at shipper's option. When jointed, veneers reasonably matched for colour. Small defects and sound knots admitted. Occasional open splits well filled. No worm holes.

Reverse jointed or unjointed at shipper's option. When jointed, veneers unmatched for colour. Occasional small open joints admitted. No worm holes.

Note

i When quoting size the length, i.e. the dimension along the grain of the face veneer is given first. When this dimension is smaller than the width, i.e. across the grain, then the board is said to be cross grain.

ii The number of plies present will depend on the thickness and manufacturer in British made plywood it may vary from 3 plies at 5 mm to 9 plies at 25 mm.

Plywood species, sizes, thicknesses, grades and adhesives

Species	Producing Countries	Main sizes in mm	Thicknesses in mm	Grades	Adhesives
Alder	Czechoslovakia	1830/2200 x 1220 1220 x 1830/2200	4–6	BB/BB BB	Extended resin
	Russia	1525 x 1220 1425 x 1425 1475 x 1475 1270 x 1270 1525 x 1525	3–6	BB/BB BB BB/C C (C unsanded)	Casein/ Blood Albumen
Ash (brown) see also Tamo & Sen	China	1830/2135 x 915/1220	3–18	1st, 2nd & 3rd	Extended resin MR and WBP
Baboen (Dalli)	Dutch Guiana	2440 x 1220	4–18	A/B A/C B/BB BB B/C	Extended resin MR
Basswood See also Shina	China	1830/2135 x 915 1525/1830/2135 x 1525 1830/2135/2440 x 1220	3–18	1st, 2nd & 3rds	Casein Extended resin Some MR and WBP
Beech	Austria	up to 2440 x 1525	4–12	A/B B/BB BB BB/BBB	Casein Extended resin and MR
	Bulgaria	2010 x 1250 1250 x 2007 Flush door sizes	3–12	B BB and C and combinations of them	Extended resin
	Czechoslovakia	1830/2200 x 1220 1220 x 1830/2200	4–12	B B/BB BB	Extended resin

Plywood species, sizes, thicknesses, grades and adhesives (continued)

Species	Producing Countries	Main sizes in mm	Thicknesses in mm	Grades	Adhesives
	Denmark	up to 2135 x 1220	4–25	B B/BB BB	Extended resin MR and WBP
	Germany	up to 2440 x 1535	8–25	A/B (I/II) B (II) B/BB (II/III) C (IV)	Extended resin MR and WBP
	Roumania	1830 x 1220 1220 x 1830 2010 x 1245 1245 x 2010 1540 x 1540 2200 x 1220 1220 x 2200 2440 x 1220 Flush door sizes	3–15	B B/BB BB Standard grade	Casein and Extended resin
	Yugoslavia	up to 2490 x 1525 and Flush door sizes	3–12	A/BB AJ/BB B B/BB BJ/BB BB BB/WG	Casein and extended resin. Some MR
	Japan	1830/2440 x 915/ 1220 and Flush door sizes	3–9	1st and 2nd	Extended resin. Some MR and BR

Plywood species, sizes, thicknesses, grades and adhesives (continued)

Species		Sizes in mm	No of plies	Thickness (mm) NOMINAL	Actual Unsanded	Sanded	Grades	Adhesives
Birch, European	Finland	**Standard Imperial**						Exterior
		1220 x 1220/1830/2135/	3 ply	4	4.2	3.8	B (I)	
		2440/2745/3050/3660	5 "	6.5	7.0	6.5	B/BB(I/III)	Extended
			7 "	9.0	9.8	9.3	B/WG(I/IV)	Resin
		Standard Metric	9 "	12.0	12.6	12.0	S (II)	WBP
		1200 x 1200/1800/2100/	11 "	15.0	15.4	14.8	S/BB(II/III)	
		2400/3000/3600	13 "	18.0	18.2	17.6	S (WG(II/IV)	
		1500 x 1500/1800/2100/	15 "	21.0	21.0	20.4	BB (III)	
		2400/2700/3000/3600	17 "	24.0	23.8	23.2	BB/WG(III/IV)	
							WG (IV)	
			TOLERANCES RANGE FROM ±0.3 – ±0.5 mm					
		Plywood mainly for concrete shuttering resin bonded and scarf jointed.						
		Standard Imperial						
		A range of Panels up to 1525 x 3050 and 3050 x 1525		4, 6.5, 9 12, 15, 18, 21, 24				
		Metric Sizes						
		Up to 1500 x 3000 and 3000 x 1500						
		Extra Large Sizes						
		Width up to 3660 Lengths according to requirement Joints covered by film overlay.						All edges are factory coated and sealed with water repellent protectives

Plywood species, sizes, thicknesses, grades and adhesives (continued)

Species	Producing Countries	Main sizes in mm	Thicknesses in mm	Grades	Adhesives
Birch, European	Russia	1270 x 1270, 1525 x 1270	3	BB/CP	
		1270 x 1220			
		1525 x 1525, 1475 x 1475	4	BB	
		1650 x 1525			
		1525 x 1525, 1270 x 1525	6	BB	
		1220 x 1220		BB/CP	
		1525 x 1525	8	B/BB	
		1525 x 1525, 1270 x 1270/			
		1475 x 1475		B/CP	
		1525 x 1525		BB	
		1270 x 1270	9	BB ext WBP	
		1270 x 1525		BB ext WBP	
		1525 x 1525, 1475 x 1475		BB/C	
		1525 x 1525, 1270 x 1270		BB	
		1525 x 1525, 1270 x 1270		BB/CP	Casein
		1525 x 1475, 1475 x 1525			Blood Albumen
		1475 x 1475, 1525 x 1475	12		Extended resin
		1475 x 1525			and WBP
		1525 x 1525		BB/C	
		1525 x 1475, 1475 x 1525		BB/CP	
		1475 x 1475, 1525 x 1270	15		
		1525 x 1525, 1475 x 1525			
		1475 x 1475		BB/C	
		1270 x 1525, 1270 x 1270		BB	
		1525 x 1525, 1270 x 1270			
		1525 x 1475, 1475 x 1525	18	BB/CP	
		1475 x 1475			
		1525 x 1270, 1425 x 1525			

Plywood species, sizes, thicknesses, grades and adhesives (continued)

Douglas Fir	Canada	**Standard** 3050/2440/2135/1830 x 1525/1220 **Metric** 2400 x 1200	6.5–32	Good two sides Good one side Good solid Solid two sides Solid one side Regular & select Sheathing Overlays	WBP No interior glue line.
	U.S.A.	2440 x 1220 3660/1830 x 1525/1220	6.5–25	Interior A/A A/B A/D B/B B/D C/D .Exterior A/A A/B A/C B/B B/C C/C	Casein Soya and extended resin WBP
Elm, Japanese	Japan	1830 x 915/1220 and Flush door sizes	3–6	1st and 2nd	Extended resin Some MR
Fuma	Belgium	up to 2440 x 1220	3–25	B B/BB	Extended resin Some MR BR and WBP
Gaboon or Okoume	England	up to 3050 x 1830	3–25 & thicker	1/1, 1/2, 2/2, 2/3, 3/3	MR and WBP
	France	up to 3050 x 1525	3–25	A/B A/BB B B/BB BB	Extended resin MR BR and WBP
	Germany	up to 3050 x 1525	3–25	1/11 1/111 11 11/111 111	Extended resin MR BR and WBP
	Holland	up to 3050 x 1525	4–18	A/B A/BB A/C A/D B B/BB B/C B/D	MR and WBP

Plywood species, sizes, thicknesses, grades and adhesives (continued)

Species	Producing Countries	Main sizes in mm	Thicknesses in mm	Grades	Adhesives
	Spain	up to 2440 x 1525	3–25	A/B B B/BB	Extended resin MR and BR
	Norway	up to 2440 x 1220	3–25 •	A/B A/BB B B/BB BB	Extended resin MR and BR
	Gabon (West Africa)	up to 2440 x 1220 up to 2135 x 1525	4–25	Premier & Standard	MR and AX100
	Israel	up to 2440 x 1525 up to 3050 x 1220 Flush door sizes	3 and thicker	B B/BB BB	WBP
Ilomba	England	up to 2440 x 1220	2–12	1/2, 2/2, 2/3	WBP
	Spain	up to 2440 x 1525	3–25	B B/BB	Extended resin, MR and WBP
	Norway	up to 2440 x 1220	4–25	II, III, IV	Extended resin MR and WBP
Lauan	Japan	1830 x 915/1220 and Flush door sizes	3–18	1st, 2nd, 3rd	Extended resin MR and BR
	China	1830 x 915/1220 and Flush door sizes	3–18	1st, 2nd, 3rd	Extended resin MR
Limba/afara	Belgium Congo Republic France Holland Germany Portugal	up to 2440 x 1525 and Flush door sizes	4–18	A/B B B/BB I/II, II, II/III	Extended resin MR and WBP
	Norway	up to 2440 x 1220 and Flush door sizes	3–25	A/B A/BB B B/BB BB	Extended resin MR and BR

Plywood species, sizes, thicknesses, grades and adhesives (continued)

Mahogany (African Mahogany)	England	up to 3050 x 1830	4–25	1/1 1/2 2/2 2/3 3/3	MR and WBP
	Nigeria	up to 2440 x 1220 and Flush door sizes	4–18	Gold Silver Blue	WBP
	Holland	up to 3050 x 1525	4–25	A/B A/BB A/C A/D B B/BB B/C B/D	MR and WBP
	Ghana	1830/2135 and 2440 x 1525	4–25	1st, 2nd, 3rd	MR
		2440 x 1220 and Flush door sizes	4–18	Gold Silver Blue	WBP
	Israel	up to 2440 x 1525 up to 3050 x 1220 and Flush door sizes	3 and thicker	B/BB B	WBP
	France	up to 3050 x 1525	3–25	A/BB B B/BB	MR, BR and WBP
	Austria	up to 2440 x 1220	4–25	1st and 2nd	Extended resin
	Italy	up to 3050 x 1525	3–25	1st and 2nd	Extended resin
	Norway	up to 2440 x 1220 and Flush door sizes	3–25	A/B A/BB B B/BB BB	Extended resin MR and BR
	Spain	up to 2440 x 1525	3–25	B B/BB	Extended resin MR and BR
Makore	England	up to 3050 x 1830	3–25	1/1 1/2 2/2 2/3 3/3	MR and WBP
	Holland	up to 3050 x 1525	4–25	A/B A/BB A/C A/D B B/BB B/C B/D	MR and WBP
	Ghana	2440 x 1220 Flush door sizes	4–18	Gold Silver Blue	WBP
	France	up to 2440 x 1525	4–25	A/BB B B/BB	MR and WBP

Plywood species, sizes, thicknesses, grades and adhesives (continued)

Species	Producing Countries	Main sizes in mm	Thicknesses in mm	Grades	Adhesives
Mengkulang	Malaya	2440 x 1220	3–25	B B/BB	WBP
Oak European (sliced cut, figured or plain used as face veneers)	England	up to 2440 x 1525	4–18	1/1 1/2 2/2 2/3 3/3	Extended resin MR and WBP
	Austria Belgium France Germany Holland Italy	up to 2440 x 1525	3–18	1st and 2nd	Extended resin
Oak Japanese figured & plain	Japan	1830/2135/2440 x 915/1220 and Flush door sizes	3–9	1st and 2nd	Extended resin, MR
Rotary cut	Japan	1830/2135 x 915/1220	3–9	1st and 2nd	Extended resin, MR
Obeche including Wawa and Samba	England	up to 2440 x 1525	4 and up	2/2 2/3 3/3	MR and WBP
	France	up to 2440 x 1525	3–25	B B/BB A/B	MR and WBP
	Germany	up to 3050 x 1525	3–25	I/II II II/III	MR and WBP
	Italy	up to 2240 x 1525	3–25	B B/BB	Extended resin MR and WBP
Pine, Columbian or Oregon	See Douglas fir				
Pine, European	Czechoslovakia	1830 x 1220 1220 x 1830	4–6	B/BB BB	Extended resin

Plywood species, sizes, thicknesses, grades and adhesives (continued)

Species	Origin	Sizes	Thicknesses	Grades	Adhesives
	Norway	up to 2440 x 1220	1–18	A/B A/BB B B/BB BB	Extended resin. Some MR and WBP
	Sweden	1270 x 1270 1525 x 1525 1830 x 1270 2032 x 1016 2135/2440 x 1220	4–18	A/B B B/BB BB A/X B/X BB/X	Extended resin MR and WBP
	Poland	up to 1060 x 1220	4–18	A/B B B/BB BB	Casein
Wet cemented	Poland	1550 x 1245 1060 x 1245	3 and 4	BB B/BB BB/BB	Casein
Pine, Chile	Chile	2185 x 1220/1525 2440 x 1220/1525 1525 x 1220/1525	4–22 3–22	1 Crown 2 Crowns 1 Star 2 Stars	Casein BR WBP
Pine, Parana or Araucaria	Brazil	1600 x 1600 2210 x 1600 up to 2490 x 1220/1600	3 and up	A/B B B/BB	Casein MR and WBP
Poplar	Italy	up to 2440 x 1525	3–12	A/B B/BB BB 1st and 2nd	Extended resin Some MR and WBP
	Canada	2440 x 1220	6.5–25	Good one side Good two sides Good/Solid Solid one side Solid two sides	WBP
Sapele rotary cut	As for Mahogany				
sliced cut	As for European Oak				

Plywood species, sizes, thicknesses, grades and adhesives (continued)

Species	Producing Countries	Main sizes in mm	Thicknesses in mm	Grades	Adhesives
Sen or White Ash	Japan	1830/2135 x 915/1220 and Flush door sizes	3 and up	1st and 2nd	Extended resin MR
Seraya	England	1830 x 915/1220 2440 x 1220 and up to 3050 x 1830	3–25	1/1 1/2 2/2 2/3 3/3	MR and WBP
Shina (Japanese Lime) or Basswood	Japan	1830/2135 x 915/1220 and Flush door sizes	3 and up	1st and 2nd	Extended resin MR
Sterculia (Red and White)	Nigeria	up to 2440 x 1220 and Flush door sizes	4–18	Gold Silver Blue	WBP
Tamo (Japanese Ash or Brown Ash)	Japan	1830 x 915/1220 2135 x 1220	3–12	1st and 2nd	Extended resin and MR
Tola Branca	Portugal	up to 2440 x 1220	3–18	I II II/III III/IV	MR and WBP
	Germany	up to 2440 x 1525			
	England	up to 2440 x 1220	4–25	II/II	MR and WBP
Utile	As for Mahogany	2440 x 1220 and Door sizes			
Virola		2440 x 1220	4–18	B/BB	WBP
Yang	China	1830/2135/2440 x 915/1220 and Flush door sizes	3 and up	1st, 2nd, 3rd	Extended resin, MR and WBP
	Thailand	up to 2440 x 1220	3 and up	I/I to III/IV BB/C	Extended resin MR and WBP

Appendix F

Blockboard and laminboard: quality and sizes.

Table 1. Blockboard species, and sizes

Faces	Cores	Producing Countries		Sizes in mm	Thicknesses	Adhesives
Beech, European	Pine Spruce Fir	Belgium		up to 1525 x 3050	16 – 25	Extended resin
		Czechoslovakia		up to 1220 x 2440	16 – 25	Casein and extended resin
		Germany		up to 1830 x 5105	13 – 25	Extended resin MR & BR
		Roumania		up to 1220 x 2010	18	Casein & extended resin
		Denmark		1220 1220 x 2400	13 – 25	Extended resin
		Austria		1220 x 2400	16 – 25	Casein & extended resin
Birch, European	Pine	Finland	5 Ply	**Standard Imperial** 1220/1525 x 1220/ 1525/ 1830/ 2135/ 2440/ 2745/ 3050/ 3660 **Metric** 1200/1500 x 1200 1500 1800 2100 2400 3000 3600	12 – 25	Extended resin
			3 Ply	As Above	12 – 25	
		Czechoslovakia		1220 x 2440	16 – 25	Casein & extended resin

Table 1. Blockboard species, and sizes. (continued)

Faces	Cores	Producing Countries	Sizes in mm	Thicknesses	Adhesives
Gaboon	Pine Spruce Fir	Austria	up to 1830 x 3480	16 – 25	Extended resin
		Germany	up to 2135 x 5105	13 – 38	Extended resin MR & BR
		Czechoslovakia	2440 x 1220	18	
		Holland	up to 1525 x 4495	15 – 25	Extended resin MR, BR
	Obeche/Pine	Israel	1525 x 2440	16 – 25	BR
	Pine	Belgium	up to 1220 x 2440	12 – 25	Extended resin
	Pine Western, Red Cedar W. African woods	England	up to 1830 x 3660	16 and up	MR and BR
	Poplar/Pine/Gaboon	France	up to 1525 x 3300	16 – 25	Extended resin MR
Ilomba, Limba/Afara Mahogany(W. Africa) Makore, Obeche	Poplar Pine and W. African woods .	England Germany Holland	as for gaboon		
Sterculia Gedu Nohor	W. African woods	Ghana Nigeria	up to 1220 x 2440	18	MR
Poplar	Pine and Poplar	Belgium	up to 1525 x 4495	13 – 25	Extended resin
		France	up to 1525 x 4570	16 – 25	Extended resin
		Italy	up to 1525 x 4495	16 – 25	Casein & extended resin
Pterygota	W. African woods	Nigeria	up to 1220 x 2440	18	MR and some WBP
Seraya	Pine, W. Red Cedar W. African woods	England	up to 1830 x 3660	16 and up	MR and BR

Table 1. Blockboard species, and sizes. (continued)

Tola Branca (Agba)	Pine	Holland Portugal Germany	up to 2020 x 4495	13 – 25	Extended resin

Table 2. Laminboards, species and sizes.

Faces	Cores	Producing Countries	Sizes in mm	Thicknesses in mm	Adhesives
Birch, European	Pine/Birch	Finland	up to 1220 x 3660	12 – 25	Extended resin
Fuma (Ceiba)	Fuma	Congo Republic	up to 1525 x 3050	16 – 30	Extended resin
Gaboon	Gaboon/Pine Spruce	Germany	up to 1830 x 5105	13 – 25	Extended resin MR & BR
	Fir/Obeche	Holland	1525 x 4495	13 – 25	Extended resin MR & WBP
	Pine/Western red cedar/W. African woods	England	up to 1830 x 3660	16 and up	MR and BR
Ilomba	W. African woods	England	up to 1830 x 3660	16 – 30	MR and BR
Limba, Mahogany etc.	W. African woods	Holland	up to 1525 x 4495	15 – 25	Extended resin MR & WBP
	Fir/Spruce/Pine & W. African woods	Germany	1830 x 5105	13 – 25	Extended resin MR & BR
	Fuma	Congo Republic	up to 1525 x 3050	16 – 30	Extended resin
Tola Branca (Agba)	Fuma	Congo Republic	up to 1525 x 3050	16 – 18	Extended resin
W. African Woods	W. African woods	Nigeria	1220 x 2440	18	MR

Note that both blockboard and laminboard are obtainable in decorative veneers including oak, teak, afrormosia and sapele.

Appendix G

Durability of timber species

Treatment and application: external walls

Introduction

This appendix includes:

1 A list of commercially available timbers, suitable for cladding, arranged alphabetically within different categories of durability. Indication is given of inherent stability, with regard to moisture movement, and of the supply position at time of writing. In a third column, species are classified according to distinction between hardwood and softwood. Lastly, under the columns headed 'E & W' and 'S' are shown those species listed in regulation B.3 (as amended) of *The Building Regulations* 1965 (covering England and Wales) and, in Schedule 10 of *The Building Standards* (Scotland) (Consolidation) Regulations 1970 respectively. These two sets of regulations permit certain species to be used untreated and require other species of less durable qualities to be preservative treated. In these last two columns the letter 'D' signifies that the species may be used in its natural state and the letter 'T' indicates the species must be preservative treated to comply with the appropriate Regulations.

2 A summary of treatments given in CP.98: 1964 – *Preservative Treatments for Constructional Timber.*

3 TRADA recommendations for relating natural durability treatment and use.

Durability of Commercial Timbers suitable for Exterior Cladding.

Non-Durable (Perishable)

Commercial Name	*Botanical Name*	*Stability* (Moisture movement)	*Supply*	*Class*	*E & W*	*S*
Abura	*Mitragyna ciliata*	S	G	H	T	–
Afara	*Terminalia superba*	S	G	H		–
Avodire	*Turraeanthus africanus*	S	G	H		–
Elm, European	*Ulmus procera*	M	G	H	T	–
Elm, Wych	*Ulmus glabra*	M	F	H	T	–
Seraya, White	*Parashorea spp*	S	G	H		–
Seraya, Yellow	*Parashorea spp*	M	G	H		–
Seraya, Red	*Parashorea spp*	S	G	H		–

Durability of Commercial Timbers suitable for Exterior Cladding. (continued).

Non-Durable (Perishable) *Commercial Name*	*Botanical Name*	*Stability (Moisture movement)*	*Supply*	*Class*	*E&W*	*S*
Hemlock, Western	*Tsuga heterophylla*	M	G	S	T	T
Pine, Scots (Redwood)	*Pinus silvestris*	M	G	S	T	T
Spruce, European (Whitewood)	*Picea abies*	M	G	S	T	T
Spruce, Canadian (Western White)	*Picea glauca var. albertiana*	M	G	S		T
Spruce, Canadian (Eastern)	*Picea glauca*	M	G	S		T
Spruce, Sitka	*Picea sitchensis*	M	G	S	T	T
Moderately Durable						
‡Ayan	*Distemonanthus benthamianus*	S	F	H		–
Danta	*Nesogordonia papaverifera*	M	F	H	D	–
†Dahoma	*Piptadenia africana*	M	F	H	D	–
Gedu nohor	*Entandrophragma angolense*	S	G	H	D	–
*Keruing/Gurjun/Yang	*Dipterocarpus spp*	L	G	H	D	–
Karri	*Eucalyptus diversicolor*	L	S	H		–
Mahogany, African	*Khaya spp*	S	G	H	D	D
Niangon	*Tarrietia utilis*	M	G	H	D	–
Oak, Turkey	*Quercus cerris*	L	S	H		–
Sapele	*Entandrophragma cylindricum*	M	G	H	D	D
Walnut, African	*Lovoa klaineana*	M	G	H		–
Fir, Douglas	*Pseudotsuga taxifolia*	S	G	S	T	T
Larch, European	*Larix decidua*	L	G	S	T	–
Larch, Japanese	*Larix leptiolepis*	L	F	S	T	–
Pine, Maritime	*Pinus pinaster*	M	G	S		–
*Pine, Pitch, American	*Pinus palustris*	M	G	S		–
*Pine, Pitch, Caribbean	*Pinus caribaea*	M	G	S		–

Durable *Commercial Name*	*Botanical Name*	*Stability (Moisture movement)*	*Supply*	*Class*		
Agba	*Gossweilerodendron balsamiferum*	S	G	H	D	D
Californian Redwood	*Sequoia (Sempervirens)*	S	S	S	D	D
Cedar, American	*Cedrela spp*	S	S	H		–
†Cedar, Western Red	*Thuja plicata*	S	G	S	D	D
Cedar, Yellow	*Chamaecyparis nootkatensis*	S	S	S		–
Cedar, Central American	*Cedrela spp*	S	S	H		–
†Chestnut, Sweet	*Castanea sativa*	S	F	H		–
Cypress (Southern or Swamp)	*Taxodium distichum*	S	S	S		–
Guarea	*Guarea cedrata*	S	G	H	D	
Kempas	*Koompassia malaccensis*	M	G	H	D	–
†Idigbo	*Terminalia ivorensis*	S	G	H	D	D
Meranti, Dark Red	*Shorea spp*	M	G	H	D	–
†Oak, American white	*Quercus spp*	M	G	H		–
†Oak, European	*Quercus/robur/petraea*	M	G	H	D	D
Utile	*Entandrophragma utile*	M	G	H	D	D
Yew	*Taxus baccata*	S	S	S		–
Very Durable						
†Afrormosia	*Afrormosia elata*	S	G	H	D	D
‡Afzelia (Apa)	*Afzelia spp*	S	G	H	D	–
Camphorwood East African	*Ocotea usambarensis*	S	F	H		–
Iroko	*Chlorophora excelsa*	S	G	H	D	D
‡Jarrah	*Eucalyptus marginata*	M	F	H		–
‡Kapur	*Dryobalanops spp*	M	S	H	D	–

Durability of Commercial Timbers suitable for Exterior Cladding. (continued).

Very Durable *Commercial Name*	*Botanical Name*	*Stability* (Moisture movement)	*Supply*	*Class*	*E&W*	*S*
Mahogany, American	*Swietenia macrophylla*	S	G	H		—
‡Makore	*Mimusops heckelii*	S	G	H	D	D
Mansonia	*Mansonia altissima*	M	G	H		—
Muninga	*Pterocarpus angolensis*	S	F	H		—
Opepe	*Sarcocephalus diderrichii*	S	G	H	D	D
Peroba white	*Paratecoma peroba*	M	S	H		—
Teak	*Tectona grandis*	S	G	H	D	D

Notation

Special Characteristics	*Stability (Moisture Movement)*	*Supply*	*Class*
*Exudes gum.	S—small	G—good	S—Softwood
†Requires care—Iron staining	M—medium	F—fair	
‡Should be pre-bored for nailing.	L—large	S—short	H—Hardwood

NB—Durability grading is based on properties of heartwood. Most commercial hardwoods can be obtained in grades without a significant proportion of sapwood being present. Most commercial softwoods from Northern Europe contain a high proportion of sapwood but recommendations at the end of this appendix allow for this.

Summary of treatments referred to in CP.98:1964.

Types of Preservative

TO1 = Tar-oil; cresote to BS 144.

TO2 = Tar-oil; other types to BS 3051.

OS1 = Organic-solvent; Chlornaphthalenes.

OS2 = Organic-solvent; Metallic naphthenates.

OS3 = Organic-solvent; Pentachlorophenol derivatives and salts.

WB1 = Water-borne; Copper/chrome to BS 3452.

WB2 = Water-borne; Copper/chrome/arsenate to BS 4072

WB3 = Water-borne; Fluoride/arsenate/chromate/dinitrophenol to BS 3453.

WB4 = Water-borne; boron compounds, sodium fluoride and sodium o-phenylphenoxide.

Methods of treatment

PI *Pressure Impregnation* to BS 913 for TO1 and TO2; to BS 3452 for WB1; to BS 3453 for WB3; BS 4072 for WB2.

HC *Hot-and-Cold Open Tank Treatment.* Timber immersed in preservative is heated to about 85°C, and after a few hours allowed to cool, drawing in preservative.

ST *Steeping.* Timber is submerged in cold preservative for a period varying from ten minutes to a few weeks, according to size, species, type of preservative and purpose. (n. b. England and Wales Building Regulations (reg. B3 as amended) requires not less than 1 hour for redwood [European] and Scots Pine).

DP *Dipping.* Comparatively dry timber (preferably well below 25% moisture content) is submerged in preservative for at least 10 seconds.

BS *Brushing and Spraying.* Generally used for in-situ work where other methods are impossible. Application must be 'liberal'. TO types of preservative should be warmed before use, especially in cold weather.

TRADA recommendations

1 Timber of any species must be free from defects, such as splits and shakes and loose knots, likely to lead to ingress of moisture.

2 All hardwoods, and also Western Red Cedar and Californian Redwood *(Sequoia)*, which are listed as DURABLE or VERY DURABLE can be used for external cladding without treatment, provided that a significant proportion of sapwood is not present.

3 Hardwoods listed as MODERATELY DURABLE may be painted or finished with clear varnish without any pre-treatment with preservative. Preservative Stains or Clear Water Repellants may also be used (see TRADA publications,'Advisory Leaflet No.2', 'Research Reports' Nos: WT/RR/6 and WT/RR/8). Alternatively such hardwoods may be preserved in accordance with CP 98 without painting (see Table 1 Group 2 of this Appendix.

4 NON-DURABLE hardwoods may also be finished with paint or clear varnish but pre-treatment with preservative in accordance CP 98 (see Table 1 Group 3 of this Appendix) is recommended to enhance the life of the cladding.

5 Softwoods, which are listed as MODERATELY DURABLE or NON-**DURABLE** all require preservative treatment to comply with the Regulations; regardless of subsequent painting in England and Wales; and if unpainted or in situations of inaccessibility, in Scotland.

It would appear reasonable to apply for relaxation of the Regulation requiring preservative treatment of softwood boarding, where the boarding is to be adequately painted and is in a position readily accessible for inspection and maintenance.

Table 1. Situations, Preservatives and Methods.
(based on Table 4, CP 98:1964)

	Situation	Preservative	Method
1	External in Ground Contact	PI	TO1, 2; WB1, 2
		HC	TO1, 2; OS2, 3; WB1, 2,
2	External NOT in ground contact and NOT painted	PI	TO1, OS3; WB1, 2, 3
		HC	TO1, 2; OS2, 3; WB1, 2, 3
		ST	TO1, 2; OS1, 2, 3
		DP, BS	TO1, 2; OS1, 2, 3
3	External NOT in Ground Contact, Treated *AND* painted	PI	WB1, 2, 3
		HC	OS2, 3; WB1, 2, 3
		ST	OS1, 2, 3; WB4
		DP, BS	OS1, 2, 3; WB4
4	Interior, if in contact with Damp Foundations	PI	TO1, OS3; WB1, 2
		HC	TO1, 2; OS2, 3; WB1, 2
		ST	TO2; OS1, 2, 3
6	House Longhorn Beetle	PI	TO1; WB1, 2, 3
		DP, BS	TO1, 2; OS1, 2, 3

Within each group treatments are given in the order of their probable effectiveness.

Appendix H

Fire treatment of timber and performance data

Surface spread of flame on timber, plywood and chipboard

Scope

This Appendix deals with the following:

1 Intrinsic flamespread resistance of timber.
2 Various treatments available for bringing timber and timber products up to Class 1.
3 The general position regarding Class O boards.
4 Note on excluded surfaces.
5 Note on the application of finishes.

1 Intrinsic flamespread resistance of timber

Knowledge of the behaviour of untreated timber is based on a series of tests carried out by the Fire Research Station, recorded in F.R.S. Note No. 553 of June, 1964.

These show that, although performance can be influenced by moisture content, thickness, surfacing and species, timber of a density greater than 400 Kg/m^3 has intrinsic Class 3 flamespread property. Western Red Cedar (and Balsa) were the only species tested which failed to meet the Class 3 criteria.

2 Treatments to achieve Class 1

Three basic methods are employed.

A *Impregnation* with a solution of chemicals, such as monammonium phosphate, ammonium chloride or borax. Timber, plywood and chipboard can be treated in this way. Advice should be obtained from specialist firms, particularly regarding compatibility with metals and finishes.

B *Surface Coatings* either in the form of a paint, paste, emulsion or clear varnish can be applied to timber, plywood and chipboard usually by brushing, but in some cases by trowel or 'curtain coating' processes.

C *Bonded Facings* of non-combustible materials such as asbestos felt, or thin metal sheets are used particularly in conjunction with chipboard.

Special grades of chipboard and plywood are made, incorporating fire retardant chemicals in their manufacture.

This Appendix shows a list of products which have passed test as Class 1, reproduced by courtesy of the Director, Fire Research Station, from more comprehensive tables of test results in Fire Note No. 9.

Products tested include both specially manufactured boards and treatments, by impregnation or surface coating, which can be applied to various timber based materials. An indication of this distinction is shown thus –
*manufactured flamespread resistant board. † impregnation treatment,
‡ applied surface coating.

For test purposes, treatments are of necessity applied to timbers, plywoods and particle boards involving a particular species or type of wood and particular thicknesses. It would be quite impractical to test them as applied to all combinations of species and thickness likely to be used in buildings. Particular test results can reasonably be regarded as applicable to treatments when applied to a range of similar species and thicknesses. In doubtful cases the Fire Research Station or appropriate manufacturers should be consulted.

Table of Tests passing Class 1 — Wood and Wood Based Materials — Timber

Firm		*Product*	*TEST No.*
Albi-Willesden Ltd., c/o Rentokil Laboratories Ltd., Wood Preserving Division, Felcourt, East Grinstead, Sussex.	‡	Softwood, 19.1 mm ($\frac{3}{4}$ in.) with two coats 'Albi-Saf No. 1', total 4.6 m^2/ℓ(25 yd^2/gal).	1511
	‡	Hardwood, 25.4 mm (1 in.) painted with 'Albi-Clear No. 1', 4.05 m^2/ℓ (22 yd^2/gal).	c/53921/3
	‡	Hardwood, 25.4 mm (1 in.) painted with 'Albi-Clear No. 1', 4.05 m^2/ℓ (22 yd^2/gal), 3078 overcoated with Chlorinated Rubber Varnish D.S.D. 10351 at 70 yd^2/gal.	Yarsley
British Paints Ltd., Brittanic Works, Portland Road, Newcastle upon Tyne, 2	‡	Softwood, 22.2 mm ($\frac{7}{8}$ in.) with two coats 'Extinite A' Fire Retarding Paint, 2.15 and 323 g/m^2 (6.36 and 9.54 oz/yd^2) respectively and two coats 'Murisan' Emulsion Paint, each at 75.1 g/m^2 (2.23 oz/yd^2).	2346
Celcure & Chemical Co. Ltd., 300 Bearsden Road, Glasgow, W.3	†	Softwood, 19.1 mm ($\frac{3}{4}$ in.) pressure impregnated with 'Celcure' Fire Retardant F(3), net salt retention 0.398 g/cm^3 (2.49 lb/ft^3).	1033
Exsud Engineering Ltd., 55 55 Aldgate High Street, London, E.C.3	‡	Softwood, 9.5 mm ($\frac{3}{8}$ in.) painted with 'Exolit Firestop' 295 g/m^2 (8.78 oz/yd^2).	971
Hicksons Timber Impregnation Co. (G.B.) Ltd., Castleford, Yorks.	†	Softwood, 38.1 mm (1$\frac{1}{2}$ in.) impregnated with 'Pyrolith', net salt retention 0.04 g/cm^3) (2.5 lb/ft^3)	1579
	†	Softwood, 25.4 mm (1 in.) impregnated with 'Pyrolith Plus' net salt retention 0.04 g/cm^3 (2.5 lb/ft^3).	2460
Thomas Hinshelwood & Co. Ltd., Glenpark Street, Glasgow, E.1	‡	Softwood, 12.7 mm ($\frac{1}{2}$ in.) painted with 'Frijol' Fire Retarding Paint, 530 g/m^2 (15.75 oz/yd^2)	1335
Hoben Davis Ltd., Spencroft Road, Holditch Industrial Estate, Newcastle-under-Lyme	‡	Timber, 19.1 mm ($\frac{3}{4}$ in.) coated with 'Fillerseal', 2160 g/m^2 (63.9 oz/yd^2).	1258
The Marbolith Flooring Co. Ltd., 3 Corbetts Passage, London, S.E.16	‡	Timber board, 19.1 mm ($\frac{3}{4}$ in.) with one coat 'Marbolith' 19.1 mm ($\frac{3}{4}$ in.)	1250
Mellor Mineral Mills Ltd., Etruria Vale, Stoke-on-Trent	‡	Douglas Fir, 12.7 mm ($\frac{1}{2}$ in.) painted with 'Dohm Fireseal', 860 g/m^2 (25.4 oz/yd^2).	1063

North British Chemical Co. Ltd., Droylsden, Manchester	‡	Columbia Pine, 19.1 mm ($\frac{3}{4}$ in.) with one coat lead primer and two coats 'Norbritol Fire Retarding Paint No. 3458', total wet weight 618 g/m^2 (18.3 oz/yd^2).	927
Pearl Paints Ltd., Treforest Industrial Estate, Pontypridd, Glam.	‡	Softwood, 12.7 mm ($\frac{1}{2}$ in.) painted with 'Trerock' Intumat Flame Retardant Paint (E.766), 218 g/m^2 (6.5 oz/yd^2))	2314
The Timber Fireproofing Co. Ltd., Carpenter's Road, Stratford, London, E.15	†	Softwood, 19.1 mm ($\frac{3}{4}$ in.) impregnated by 'Oxylene' process, net salt retention 0.04 g/cm^3 (2.5 lb/ft^3).	FR 71 (BRS)
Vitretex (England) Ltd., 25a Castlereagh Street, London, W.1	‡	Softwood, 19.1 mm ($\frac{3}{4}$ in.) with two coats 'Snuff' Intumescent Fire Retardant Coating (FR/1), total 2.94 m^2/1 (16 yd^2/gal).	2939
Rentokil Labs Ltd., Wood Preserving Division, Felcourt, East Grinstead, Sussex.	‡	Softwood, 22.2 mm ($\frac{7}{8}$ in.) with one coat 'Pyromors-Special' wet film weight 498 g/m^2 (14.7 oz/yd^2).	1485

Table of Tests passing Class 1 — Wood and Wood Based Materials — Plywood

Firm		*Product*	*TEST No.*
Albi-Willesden Ltd.,	‡	Douglas Fir plywood, 7.9 mm ($\frac{5}{16}$ in.) with one coat 'Albi R', 247 g/m^2 (7.32 oz/yd^2).	231 (c)
c/o Rentokil Labs. Ltd., Wood Preserving Div., Felcourt, East Grinstead, Sussex	‡	Gaboon plywood, 7.9 mm ($\frac{5}{16}$ in.) with one coat 'Albi R', 247 g/m^2 (7.32 oz/yd^2).	231 (d)
		Plywood, 6 mm ($\frac{1}{4}$ in.) with two coats total 6.6 m^2 (25 yd^2/gal).	c/53921/3 Yarsley
British Paints Ltd., Brittanic Works, Portland Road, Newcastle upon Tyne, 2	‡	Douglas Fir plywood, 6.4 mm ($\frac{1}{4}$ in.) with two coats 'Extinite C' Fire Retarding Paint, total 426 g/m^2 (12.6 oz/yd^2).	2349
Hangers Paints Ltd., Stoneferry Works, Hull	‡	Gaboon-faced plywood, 6.4 mm ($\frac{1}{4}$ in.) treated with 1.6 mm (1/20 in.) coat 'Dinaphon V 103', applied by trowel, 3,780 g/m^2 (112 oz/yd^2).	952 (a)
	‡	Douglas Fir plywood, 6.4 mm ($\frac{1}{4}$ in.) treated with 1.6 mm (1/20 in.) coat 'Dinaphon V 103', applied by trowel, 3,780 g/m^2 (112 oz/yd^2).	952 (b)

Table of Tests passing Class 1 — Wood and Wood Based Materials — Plywood (continued).

Company		Material	Test No.
Rudolf Hensel Chem. Farben-und Lackfabrik, Suderstrasse 235, Hamburg 26, W. Germany	‡	Plywood, 7.9 mm ($\frac{5}{16}$ in.) with one coat 'FS 900/2 KS', 400 g/m^2 (11.8 oz/yd^2) and one coat 'FS 900/34b' 250 g/m^2 (7.4 oz/yd^2).	2826
L. T. Lewis and Co., 37, Forest View, London, E.4.		*Mahogany veneer 0.4 m/m thick bonded to 'Vedex' F.R.C.W. base without any 4305 additional treatment to the veneer.	
Lumber Products Ltd., 120, Bishopsgate, London, E.C.2.		†Veneered plywood impregnated by the Oxylene Impregnation Process. 18.08/519 (G. J. T. Crafer and Ptns).	
Permoglaze Ltd., Tyseley, Birmingham, 11	‡	Plywood, 4.8 mm ($\frac{3}{16}$ in.) with two coats 'Fyrexo' White Intumescent Emulsion Paint (T. 1791B), each at 5.15 m^2/ℓ (28 yd^2/gal).	2495
Sherwoods Paints Ltd., Barking, Essex	‡	Gaboon plywood, 6.4 mm ($\frac{1}{4}$ in.), with two coats 'Sherwoods White Fire Retardant Paint (P.L. 5069), wet film total 948 g/m^2 (28.0 oz/yd^2).	684
Stanley Smith & Co. Ltd., Worple Road, Isleworth, Middx.	*	'Ignicide B.S. Flameproof Plywood', 3.2 mm ($\frac{1}{8}$ in.) manufactured from Russian Birch.	229(a)
	*	'Ignicide B.S. Flameproof Plywood', 3.2 mm ($\frac{1}{8}$ in.) manufactured from Czech Alder.	229(b)
	*	'Ignicide B.S. Flameproof Plywood', 3.2 mm ($\frac{1}{8}$ in.) manufactured from Finnish Birch.	229(c)
	*	'Ignicide B.S. Flameproof Plywood', 3.2 mm ($\frac{1}{8}$ in.) manufactured from Gaboon.	229(d)
The Timber Fireproofing Co. Ltd., Carpenter's Road, Stratford, London, E.15	†	Gaboon plywood, 6.4 mm ($\frac{1}{4}$ in.) pressure impregnated with modified 'Oxylene' process, net salt retention 0.04 g/cm^3 (2.5 lb/ft^3).	2539
Zist Ltd., 229 Old Ford Road, London, E.3	†	Plywood, 4 mm (0.16 in.) pressure impregnated with a mon-ammonium phosphate compound, net salt retention 12 per cent.	1006

Table of Tests passing Class 1 — Wood and Wood Based Materials — Particle Board, including Chipboard

Firm		*Product*	*TEST No.*
Airscrew-Weyroc Co. Ltd., Weybridge, Surrey	*	'Weyroc' 12.7mm ($\frac{1}{2}$ in.) with asbestos felt, 0.79 mm ($\frac{1}{32}$ in.) bonded to one face.	1154
	*	'Weyroc' 19.1 mm ($\frac{3}{4}$ in.) with exfoliated vermiculite 188 915 g/m^2 (27.0 oz/yd^2) bonded to one face.	1595
Applied Acoustics Ltd., 76 Wimpole Street, London, W.1	*	'Micropor' Acoustic Panel No. 1436', 19 mm ($\frac{3}{4}$ in.) consisting of porous chipboard faced with lacquered foil of glass fibre/asbestos, 0.7 mm (0.027 in.), the whole containing fire-retardant salts (net salt retention 10 per cent by weight of panel).	2848
Borax Consolidated Ltd., Borax House, Carlisle Place, London, S.W.1; and British Plimber Ltd.	*	'Plimber Special Grade' Chipboard, 12.7 mm ($\frac{1}{2}$ in.) treated with boric acid, net salt retention 10 per cent.	2189
British Plimber Ltd., Dovers Corner, New Road, Rainham, Essex.	*	'Plimberite' Special Grade, 19.1 mm ($\frac{3}{4}$ in.) consisting of softwood shavings bound by urea-formaldehyde resin with fire-retardant additive.	992
Glikston Building Materials, Sole U.K. Concessionaires, Carpenters Road, London, E.15	*	'Glinex Board Type 400', 25.4 mm (1 in.) faced with asbestos felt, 145 g/m^2 (4.3 oz/yd^2) bonded without an adhesive.	2996
	*	'Glinex Board Type 400 A.F.', 25.4 mm (1 in.) painted with fire-retardant paint, 500 g/m^2 (14.8 oz/yd^2).	2997
	*	'Glinex Board Type 400', 25.4 mm (1 in.) painted with 'Oxylene P.V.A. Fire Retardant Emulsion Paint', 7.35 m^2/ℓ (40 yd^2/gal).	2999
Pearl Paints Ltd., Treforest Industrial Estate, Pontypridd, Glam.	‡	Chipboard, 12.7 mm ($\frac{1}{2}$ in.) with one coat 'Trerock' Flame Retardant Paint, White, 7.36 m^2/1 (40 yd^2/gal) and one coat 'Trerock' Intumat Flame Retardant Paint, White, 3.68 m^2/1 (20 yd^2/gal).	3449

*Manufactured flamespread resistant board. †Impregnation treatment.
‡Applied surface coating.

3 Class O Boards

These are defined in the Regulations (1965) as follows:

The material shall –

i be non-combustible throughout; or

ii comprise a base or *background* which is *non-combustible* with the addition of a surface not exceeding .8 mm *thick* so that the spread of flame rating of the combined product is not lower than Class 1 in clause 7 of B.S. 476: Part I:1953, or

iii comprise a base or *background* which is *combustible* but with any exposed face finished with a layer not less than 3 mm *thick* of *non-combustible material* so that the spread of flame rating of the combined product is not lower than Class 1 in clause 7 of B.S. 476: Part I:1953, and with the other face not exposed to air.

Boards can readily be manufactured to order to comply with Specification (iii). Insufficient data is available at present to prepare a list of Class 'O' boards manufactured to Specification (ii). It is hoped that it will be possible to do this in due course.

4 Excluded surfaces

Restriction of spread of flame over surfaces is applied in the Regulations *only* to internal surface of walls, ceilings, and soffits. Control is *not* applied to 'any door, door frame, window, window frame, fireplace surround, mantel shelf, fitted furniture, or *trim'* the latter being defined as 'any architrave, cover mould, picture rail, skirting or other similar member'. (Reg. E. 14(4)).

5 Application of finishes

Since the Building Regulations (1965) control only materials of which a building is constructed, it can be held that they do *not* control finishes applied subsequent to erection.

Such an interpretation might conflict with the application of fire retardant surface finishes in-situ, to meet the spread of flame requirements of the Regulations.

However, fire retardant finishes applied *before* erection would probably be accepted, provided they reach the necessary standards.

Appendix I

Sizes of particle boards

Table 1. Wood Chipboard

Normal range of common Press Sizes (mm)	Thickness (mm)	Weight (Kg/m^3)
Apart from 1220 x 2440 (48" x 96") the other common sizes available are:		Range:
1220 x 1830–2130–2740–3040–3660	3 mm – 40 mm, flat pressed boards.	350 to 720 but principally 600
1270 x 2690		
1490 x 3500		
1520 x 2740–3500–3660	16 mm – 50 mm, extruded boards	
1720 x 2740–3660		
1750 x 3500		
1800 x 2740–3660		
1830 x 2440–2740–3660		
1890 x 3590 and up		
Special sizes, including cut sizes, can be obtained on application.		

Table 2. Flaxboards

1220 x 2440	6 mm – 50 mm	Range:
1540 x 1700–1830–3500–3660		350 to 600
1830 x 7000		
Special sizes, including cut sizes, can be obtained on application.		

Appendix J

Conversion table to metric dimensions

1 inch	equals	25.4 mm	LENGTH
1 foot	"	304.8 mm	
3.28 ft	"	1.0 m	
1 in^2	equals	645.2 mm^2	AREA
1 ft^2	"	.093 m^2	
10.76 ft^2	"	1.0 m^2	
1 yd^2	"	.836 m^2	
1.2 yd^2	"	1.0 m^2	
1 ft^3	equals	.0282 m^3	VOLUME
1 std	"	4.65 m^3	
.214 std	"	1 m^3	
35.5 ft^3	"	1 m^3	
1 lb	equals	.454 Kg	WEIGHT
2.2 lbs	"	1.0 Kg	
1 lb/ft^3	"	16 Kg/m^3	
.0624 lb/ft^3	"	1 Kg/m^3	
K value 1 BTU $in/ft^2\ h\ ^\circ F$	equals	0.1442 $W/m^\circ C$	THERMAL INSULATION
6.93 "	"	1 $W/m^\circ C$	
U value 1 BTU $/ft^2\ h^\circ F$	equals	5.678 $W/m^\circ C$	
0.176 "	"	1 $W/m^\circ C$	

*These conversion factors are precise and do not necessarily agree with the actual conversions used in this publication which are either to British Standard, or rounded off to the nearest millimetre.

List of TRADA offices

Head Office and Research Laboratories
Hughenden Valley,
High Wycombe, Buckinghamshire.
Tel. Naphill 3091 (0240-24-3091)

Regional Offices

LONDON
TRADA, The Building Centre, 26, Store Street, London WC1E 7BU
Tel. 01-636 8761

MIDLANDS: D. Drury, AIWSc
TRADA, Engineering and Building Centre,
Broad Street, Birmingham, B1 2BD
021-643 1914

SOUTH WEST: R. L. Dodd,
TRADA, Bristol Building and Design Centre,
Colston Avenue, The Centre, Bristol BS1 4TW
0272 23692

WESTERN COUNTIES: R. Middleton, AIWSc
TRADA, Pearl Assurance House,
High Street, Exeter EX4 3NS
0392 54034

SCOTLAND: E. MacBride, CMIWSc. AIBICC
TRADA, The Building Centre of Scotland,
425 Sauchiehall Street, Glasgow C2
041-332 4395

NORTH EAST: D. R. Evans, AIWSc
TRADA, 18 Park Row, Leeds LS1 5JA
0532 27256

EAST: C. N. Mindham, BSc, LIOB
TRADA, 16 Trumpington Street,
Cambridge CB2 2QA
0223 66287

SOUTH EAST: W. H. Hale, AIWSc, FLS, FRMS
TRADA, The Building Centre,
26 Store Street, London WC1E 7BU
01-636 8761

NORTH WEST: H. Wilkinson, AIWSc
TRADA, The Building and Design Centre Manchester Ltd.
113–115 Portland Street, Manchester M1 6DW
061-236 3740

Select bibliography

ABBREVIATIONS USED

BWPA	British Wood Preserving Association, 62, Oxford Street, London.
FPRL	Forest Products Research Laboratory, Princes Risborough, Aylesbury, Bucks.
HMSO	Her Majesty's Stationery Office P.O. Box 569, London, S.E.1.
TDA	Timber Drying Association, Clareville House, Oxenden Street, London S.W.1.
BS	British Standard: from British Standards Institute (BSI), 1 Park Street, London W.1, or from HMSO
CP	Code of Practice: also from BSI
BRS	Building Research Station
FPDA	Finnish Plywood Development Association

GENERAL

Desch, H.E.: Timber, its structure and properties, 4th ed. 1968 Macmillan & Co £1.25

Fibre board and particle board, 1958 FAO UN (available from HMSO) £0.50

A handbook of hardwoods, 1956 HMSO for FPRL £1.50

A handbook of softwoods, 1957 HMSO for FPRL £0.62½

PLYWOOD

Plywood, revised 1966 TRADA Technical brochure

BS CP 112 The structural use of timber in buildings HMSO

Wood, Andrew, Dick and Linn: Plywoods of the world, their development, manufacture and application, revised edition 1963 W. & A. K. Johnson £2.10

TRADA Memorandum No 1

Building Regulations

MOISTURE CONTENT AND SEASONING

Timber drying, 1969 TRADA/TDA £0.10

PRESERVATION

Findlay, W. P. K.: The preservation of timber, 1962 A. & C. Black £1.25
FPRL Technical note no. 24: Preservative treatment for external joinery timber, 1968 FPRL gratis
Timber preservation, 1957 BWPA and TRADA £0.25

TIMBER AND FIRE PROTECTION

Fire and timber in modern building design, TRADA
TRADA Memorandum No 1
See 'General' above
Timber and fire protection, TRADA Red booklet series £0.40
Timber preservation, 1957 BWPA and TRADA £0.25

SURFACE TREATMENTS

Exterior clear finishes for timber, revised July 1967 TRADA advisory leaflet no 8 gratis
Floor seals, revised May 1965 TRADA advisory leaflet no 2 gratis
Gray, V. R.: Seals for timber floors Nov. 1960, TRADA research report C/RR/5
Painting woodwork, HMSO Ministry of Works advisory leaflet no 25
BRS Digest 106, 1969
Painting handbook for Finnish birch plywood, blockboard, 1968 FPDA

WOOD AND THERMAL INSULATION

BS 2972: 1961 Methods of tests for thermal insulating materials, HMSO
Computation of heat requirements for buildings – guide to common practice, 1965 Institution of Heating and Ventilating Engineers
Timber for roofs in factory buildings and the thermal insulation. Industrial building act 1957, TRADA Technical brochure

WOOD AND SOUND INSULATION

BS CP 3: 1960 chapter III, Sound insulation and noise reduction, HMSO
Data for the acoustic design of studios, 1966 BBC monograph no 64 of 1966
BS 2750: 1956 Recommendations for field and laboratory measurements of airborne and impact sound transmission in buildings, HMSO

WOOD IN BUILDING

Armstrong, F. H.: Timber for flooring, 1957 HMSO (FPRL bulletin no 40)
Eastwick-Field, J. and Stillman, J.: The design and practice of joinery, revised 1965 Architectural Press

Hardwoods for industrial flooring, 1954 HMSO (FPRL leaflet no 48)
Wood flooring, TRADA booklet
Under-floor panel heating in relation to wood block flooring, revised Feb. 1967 TRADA advisory leaflet no 1
BS CP 201: 1951 Timber flooring HMSO

NOMENCLATURE

BS 565: 1963 Glossary of terms applicable to timber plywood and joinery HMSO
BSS 881 and 589: 1955 Nomenclature of commercial timbers including sources of supply HMSO

WOOD AND WOOD-BASED MATERIALS

BS 584: 1967 Wood trim
BS 1142: 1961 Fibre building board
BS 1186 Part I: 1952 (and amendments) Quality of timber
Part II: 1955 Quality of workmanship
BS 1455: 1963 Plywood manufactured from tropical hardwoods
BS 1811: 1961 Methods of test for wood chipboards and other particle boards
BS 3493: 1962 Information about plywood
BS 2604: Part I: 1963 Resin bonded wood chipboard – imperial units
Part II: Resin bonded wood chipboard – metric units

PRESERVATION

BS 1282: 1959 Classification of wood preservatives and their methods of application
BS CP 98: 1960 Preservative treatment for constructional timber

SURFACE TREATMENTS

BS 1282: 1959 Classification of wood preservatives and their methods of application
BS 544: 1949 Linseed oil putty for use in wooden frames
BS 1215: 1945 Oil stains
BS 1336: 1946 Knotting
BSS 2521 and 2523: 1966 Lead based priming paints
BS 2524: 1966 Red oxide-linseed oil priming paint
BS 2528/32: 1954 Ready mixed oil based undercoating and finishing paints (exterior quality)
BS 2525/27: 1969 Undercoating and finishing paints for protective purposes (white-lead based)
BS 2660: 1955 Colours for building and decorative paints

ADHESIVES
BS 745: 1969 Animal glue for wood
BS 1203: 1963 Synthetic resin adhesives (phenolic and amino-plastic) for plywood
BS 1204 Part I: 1956 revised 1964 Synthetic resin adhesives (phenolic and amino-plastic) for wood. Gap filling adhesives
BS 1204 Part II: 1965 Synthetic resin adhesives (phenolic and amino-plastic) for wood. Close contact adhesives
BS 1444: 1970 Cold-setting casein adhesive powders for wood
BS 1202: Part I Steel nails
Part II 1966 Copper nails
Part III: 1962 Aluminium nails
BS 1210: 1963 Wood screws
BS 1494 Part I: 1964 Fixings for sheet, roof and wall coverings
Part II: 1967 Sundry fixings

DOORS
BS 459 Part I: 1954 Panelled and glazed wood doors
Part II: 1962 Flush doors
Part III: 1951 Fire check flush doors and wood and metal frames (half hour and one hour types)
Part IV: 1965 Matchboarded doors
BS 1567: 1953 (confirmed 1960) Wood door frames and linings
BS CP 151 Part I: 1957 Wooden doors

FLOORS: see also WOOD IN BUILDING above
BS 1187: 1959 Wood blocks for floors
BS 1297: 1970 Grading and sizing of softwood flooring
BS CP 201 Part I: 1967 Wood flooring (board, strip block, and mosaic)

STAIRS
BS 585: 1956 Wood stairs

WINDOWS
BS 644 Part I: 1951 Wood casement windows
Part II: 1958 Wood double hung sash windows
Part III: 1951 Wood double hung sash and case windows – Scottish type
BS 1285: 1966 Wood surrounds for steel windows and doors

Technical notes of the Forest Products Research Laboratory